Palgrave Studies in Animals and Literature

Various academic disciplines can now be found in the process of executing an 'animal turn', questioning the ethical and philosophical grounds of human exceptionalism by taking seriously the nonhuman animal presences that haunt the margins of history, anthropology, philosophy, sociology and literary studies. Such work is characterised by a series of broad, cross-disciplinary questions. How might we rethink and problematise the separation of the human from other animals? What are the ethical and political stakes of our relationships with other species? How might we locate and understand the agency of animals in human cultures?

This series publishes work that looks, specifically, at the implications of the 'animal turn' for the field of English Studies. Language is often thought of as the key marker of humanity's difference from other species; animals may have codes, calls or songs, but humans have a mode of communication of a wholly other order. The primary motivation is to muddy this assumption and to animalise the canons of English Literature by rethinking representations of animals and interspecies encounter. Whereas animals are conventionally read as objects of fable, allegory or metaphor (and as signs of specifically human concerns), this series significantly extends the new insights of interdisciplinary animal studies by tracing the engagement of such figuration with the material lives of animals. It examines textual cultures as variously embodying a debt to or an intimacy with animals and advances understanding of how the aesthetic engagements of literary arts have always done more than simply illustrate natural history. We publish studies of the representation of animals in literary texts from the Middle Ages to the present and with reference to the discipline's key thematic concerns, genres and critical methods. The series focuses on literary prose and poetry, while also accommodating related discussion of the full range of materials and texts and contexts (from theatre and film to fine art, journalism, the law, popular writing and other cultural ephemera) with which English studies now engages.

Josephine Taylor

The Nonhuman Narratives of Energy

Multispecies Encounters in Extraction Zones

palgrave
macmillan

Josephine Taylor
University College Dublin
Dublin, Ireland

ISSN 2634-6338 ISSN 2634-6346 (electronic)
Palgrave Studies in Animals and Literature
ISBN 978-3-031-94331-7 ISBN 978-3-031-94332-4 (eBook)
https://doi.org/10.1007/978-3-031-94332-4

Cover illustration: Ashley Cooper / Getty images

This Palgrave Macmillan imprint is published by the registered company Springer Nature Switzerland AG.
The registered company address is: Gewerbestrasse 11, 6330 Cham, Switzerland

For my Nan, Maureen Mellett, who insisted Education is no load

Acknowledgements

Vulnerability is a central modality and thread woven through this work. For me, vulnerability is the lived experience of having bipolar disorder, its unpredictability and constant state of recovery while simultaneously contending with the perils of the academic job market. In a more privileged position writing now at the time of a four-year postdoctoral fellowship, I consider the severe challenges of this book coming into being. Academic work at many times in my life provided me a sense of purpose and passion, as for me, it was always connected to social justice and agential resistance. But often it has also faltered, and so the people who surround me and support me are who this book is dedicated to.

First and foremost, I want to thank Jacob Cook, my partner for his unquestioning love, care and support over the years. The worlds we have built together, your friendship and joy are what allowed me to be inspired to care and hope for a better world for animals and our environment. You enrich my world and allow me to flourish in ways I didn't think were possible. To my family, Martin Taylor, Ann-Marie Taylor and Laura Taylor, thank you for always never questioning the dreams I want to fulfil and supporting me in everything I do. For my Nan Maureen Mellett, who has always cheered me on in my studies and academic work.

There are many people I have encountered in my academic journey who've provided friendship, mentorship and networks of support I could not do without. Laura Schack, Alex Underwood, Thomas Waterton, Edward Thornton, you have formed the inspiring foundation for this work with your care and friendship. To my wonderful friends, Roxanne Douglas, Giulia Champion, Fiona Farnsworth, Freya Verlander, Tashe

Bondre, Nora Castle, Fern Ennis, thank you for your conversations and support. Also, to the Beyond Gender Collective who supply more than just inspiration but care and solidarity.

Danielle Sands, my PhD advisor and friend, you provided the structure, creativity, care and support for this book to come to life. I thank you for everything you did in guiding me in the PhD and after. To the academic mentors and friends who have formed the bedrock of me surviving precarity: To Gavin Schwartz-Leeper, you were the best union rep I could have asked for, and since have offered continual support and friendship. To John Miller for his excellent guidance and inspiration both as an examiner and now as a collaborator. To Anat Pick for the work that you do that continually inspires me. And to Bob McKay, where the questioning of the 'human' and love for animal studies began.

Thank you to the wonderful reviewers at the contract stage and peer review process for their insightful feedback and recommendations, as well as to the series editors for their encouragement.

Competing Interests The author has no competing interests to declare that are relevant to the content of this manuscript.

Contents

List of Figures

CHAPTER 1

Introduction

Storying the Nonhuman

Extractive histories can erase the physical and narrative presence of nonhuman life. As Indigenous thinker Leanne Simpson notes in an interview with Naomi Klein,

> Colonialism and capitalism are based on extracting and assimilating. My land is seen as a resource. My relatives in the plant and animal worlds are seen as resources. My culture and knowledge is a resource. The act of extraction removes all of the relationships that give whatever is being extracted meaning. (Simpson; Klein 2013)

Extraction here names the violent mechanisms of transformation and severance, a process of erasure, which evacuates meaning and disrupts relations. We often discover that the stuff of life, biological and geological matter, is reduced to fuel and commodity in Western narratives, harnessed for the purpose of fuelling the Global North's economies. My hopes in this book are to show that there exists a peripheral nonhuman narrative in the stories and violence of extraction, hidden and erased, but often still just visible to be heard and seen. As Federici suggests, 'we [...] need a history of capitalism written from the viewpoint of the animal world' (Federici 2020, 11). This is my intention: to capture nonhuman narratives in the

J. Taylor, *The Nonhuman Narratives of Energy*, Palgrave Studies in Animals and Literature,
https://doi.org/10.1007/978-3-031-94332-4_1

grand story of energy and capitalist histories. In *The Disposition of Nature*, Jennifer Wenzel argues that 'particular literary genres, aesthetic modes, and narrative templates provide the forms through which human understandings of nonhuman nature and its dispositions are forged' (Wenzel 2020, 15). These cultural forms can reinforce the position of the nonhuman as a mere commodity, but they can also, as Wenzel puts it, relay a 'world-imagining from below', that is a form more attuned to the reality of peripheral spaces and lives (Wenzel 2020, 9). Here, I begin to sketch the storying mode which could envision from 'below', giving voice to the peripheral lives and spaces which extraction attempts to erase.

Federici's passage on the importance of the viewpoint of the animal world points to the necessity of unpacking anthropocentrism, by insisting on nonhuman life's significance to our understanding of narrative. Nonhumans are participants in the economic system of capitalism and witnesses of its violence. Storytelling and the creation of narrative demonstrates an imaginative capacity which can enable us to feel beyond the human, to invoke and experience a sense of companionship and proximity to other beings which have hitherto seemed separate, distinct and distant. Ghosh in his magnificent work *The Nutmegs Curse* uses the power of narrative to widen our understanding of how climate change is tied to the exploits of imperial resource extraction. In his analysis of the ways Western narration erases voices of particular humans and nonhuman others, the ways of life and knowing outside of rampant resource capital, he turns to Donna Haraway's invocation that 'storying cannot any longer be put into the box of human exceptionalism' (Haraway 2016, 39). Capturing the nonhuman narrative, the disruptive force of nonhumans in literature and culture, as well as their physical and material presence, allows us to contend with a conception of narrative beyond the human.

Projects of extraction from crude oil, mineral mining, natural gas, to the turning of animals' bodies into meat, often operate in periperhalised locales, severed from the immediate vision and perception of the imperial core. Narrative, the act of storytelling, whether through photography, film or fiction, can counteract how we understand nonhuman life and our wider ecologies; nonhuman narratives emerge as a disruptive force to extractive ideologies, disrupting the operations of displacement which place a clear delineation between who is consumer and who and what is a resource. I introduce the concept of nonhuman narratives of energy as a way to understand how animals feature and are a part of the history of our modern petroculture. Danielle Sands in her work *Animal Writing* suggests how nonhuman animals can be 'active participants in, and the creators of, stories, and that stories are ideal spaces for imaginative cross-species connections' (Sands 2019, 21). Storytelling, acts of creativity and the construction of alternative imaginaries allow for new paths to be forged—the specific path I wish to illuminate in this book is a path that is against extractive communities and ideologies which are perpetuated by our fossil fuel economy. My incentive here is to invite something new within our understanding of the cultural history of energy through non-human storytelling and non-anthropocentric imagining. Despite Melanie Doherty's efforts to conceive of a non-anthropocentric imagining of crude oil (Doherty 2014, 367) and Melissa Haynes questioning of why animals are so absent from energy history (Haynes 2017, 35), there has been little focus on what we can learn from a combined analysis of animal studies and energy humanities, learning the ways animals feature in histories of production and extraction, and what a nonhuman perspective might do to inform our energy and environmental futures.

To return to Ghosh's tracing of the imperial legacies and histories of resource extraction in the *Nutmegs Curse*, he considers the significance of nonhuman life and stories in accounts of imperial resource politics. Ghosh writes,

> It is perfectly possible, then, that far from being an exclusively human attribute, the narrative faculty is the most animal of human abilities, a product of one of the traits that human's indisputably share with animals and many other beings – attachments to place. Perhaps, then, storytelling, far from setting humans apart from animals, is actually the most important residue of our formerly wild selves. (Ghosh 2021, 204)

Ghosh here illuminates how narrative capacity and acts of creativity are what we share with our nonhuman companions rather than what distances us from them, storying being a shared capacity with nonhuman others. When we turn to our own stories and narratives, where does the nonhuman reside? For Timothy Baker, 'Nonhuman animals haunt the peripheries of contemporary fiction' (Baker 2019, 1). Ghosh alternatively perceives that our stories are where we can find recognition and celebration of nonhuman voices and agency. Drawing on this research of nonhuman storying, my question is: in the grand narratives of energy and capital, where do we find nonhuman life? Perhaps, like Baker suggests about contemporary fiction, they are found and haunt the peripheries. Or found in canonical works such as *Moby-Dick,* arguably an iconic early energy narrative, we then find them at the centre. To contend with the peripheral nonhuman narratives of energy is to furnish and convey a new story in the field of energy humanities. The After Oil Research collective suggest that 'the struggle that is currently taking place over the direction of energy transition, which involves scientists, activists, governments, and businesspeople, is a struggle over representation and narrative, the stories we tell about human capacity and future possibility' (After Oil 2016, 73). My contention here is that it goes beyond human capacity and representation and is to do with our nonhuman companions with whom we share our world with. If the struggle of energy transition is to do with representation and narrative, it is essential to convey the role of the nonhuman animal in our histories of energy and oil culture so that we can learn and forge a new direction away from our fossil fuel dominated economies.

Operations of Displacement and the Peripheral

Both petrocultures and animal studies have a predominant focus on vision and perception; the foundational question for these fields is: how do we capture the effects of violence and environmental destruction which are so often elided from our view? Sheena Wilson and Andrew Pendakis highlight the problematic of vision within the field of oil culture. Here they suggest,

> The problem of visualization, of the proliferation of determinate, useful maps of our economic lives, is not specific to oil, but one politically structural to a system that is at once spectacularly consumerist and fully globalized on the level of production. However, it could be argued that oil is a uniquely occluded substance: not only does its exchange value engender an enormous corporate project of hiding, an explicit machinery of deception and spin, its pervasiveness, its presence everywhere, perhaps singularly christens its position as "hidden in plain sight". (Wilson and Pendakis 2012, 4)

For Wilson and Pendakis, there is the sense that in our globalised world all commodities and their production processes are occluded. The pipeline of who is at the consumer side and the production end of resources is the question of privilege, positionality and location. They do note, however, that there is something unique about oil in its peripheralisation, the project of corporate hiding and evasion of its effects. Quoting Imre Szeman, they suggest oil is in many cases 'hidden in plain sight'. In a similar way, Maya and Reuben Fawkes in their work on *Climate Change and Art* suggest, 'oil power is revealed as operating behind the scenes' (Fowkes and Fowkes 2022, 37). Crude oil is so pervasive in consumer lifestyles in plastics and our transportation, ways of understanding contemporary life. It is embedded in our ideologies of what it means to be a mobile and liberated citizen, and it is yet sequestered away and hidden from everyday perception. Significantly, the same can said of animal products and meat, their bodies form such a significant proportion of world resources and their lives often erased from the modern consumer's gaze. In both meat and oil industries, there are specific operations of displacement and cognitive dissonance at work, a sense of erasure and problematic of vision. My contention throughout this book is to point to these erasures and offer a powerful way of viewing and feeling that can help us reconnect with the suffering we are so often disconnected and distanced from. It is key we view animal studies and petroculture in conversation in this way since both have projects of unveiling and working towards a clarity in perception, working towards an acknowledgement of the environmental destruction and animal suffering that is so often neglected.

One way of understanding these operations of displacement and geographical peripheralisation of resource commodities processes such as with oil and meat is through Marxist sociology and world systems theory. The Warwick Research Collective (WReC) draw on Immanuel Wallerstein and Leon Trotsky's formulation of combined and uneven development to illuminate contemporary capitalism machinations. Quoting Harry Harootunian, they note the centrality of sacrifice to capitalism's functioning:

> Part of the price paid for continual expansion is the production of permanent unevenness, permanent imbalance between various sectors of social formations, the process by which some areas must be sacrificed for the development of others, such as the countryside for the city [...] the colony for the metropole, or even one city for another. (Harootunian 2000, xv)

What I want to dwell on here is the notion of sacrifice, that there are indeed sacrificial zones, locations and sometimes living beings, human and otherwise, which are sacrificed in the name of global development. In the politics of crude oil, this is evidently the case. The Global North economy accrues energy and wealth from extracting in regions which are energy rich and biodiverse, areas that are occupied by Indigenous communities, as well as a variety of flora and fauna. For WReC, 'the theory of "combined and uneven development" was devised to describe a situation in which capitalist forms and relations exist alongside "archaic forms of economic life" and pre-existing social and class relations' (WReC 2015, 11). The production of the archaic and the modern is foregrounded through colonial exploitation, a process of extracting wealth and its later distribution to consumer centres, while leaving commodity productive regions bereft of the mineral resources which they produce. As WReC further note, 'capitalist development does not smooth away but rather produces unevenness, systematically and as a matter of course' (WReC 2015, 12). The capitalist regime generates what Franco Moretti describes as a system that is 'simultaneously one and unequal' (Moretti 2000, 149). Moretti's statement refers to how no regions across the world are now separate, capitalism's monopoly is global, yet the benefits accrue to the Global North while extracting from the Global South. This geographical unevenness illustrates the operations of displacement which I suggest are at work when we consider the role meat and oil production in the lives of Western consumers.

To return to Harry Harootunian's notion that capitalist development leads to a space of sacrifice, it is clear to see how the politics of oil and meat require a notion of sacrifice, whether that is a geographical area or a body; it is this erased and silenced space which I intend to focus. Such spaces of sacrifice can be described by what Macarena Gómez-Barris defines as the 'extractive zone' (Gómez-Barris 2017, xv). Extractive zone: 'refers to the colonial paradigm, worldview, and technologies that mark out regions of "high biodiversity" in order to reduce life to capitalist resource conversion' (Gómez-Barris 2017, xvi). I borrow this concept to consider not only regions under the monopoly of oil production but also how the animal's body itself is an extractive zone. This multi-layered understanding of extraction allows us to view environmental destruction and animal suffering in tandem and thus work towards an explanation of how a multispecies ethics can inform environmental futures and vice versa. In the context of Latin America, Gómez-Barris goes on to note:

> Extractivismo, as extractive capitalism is known in the Américas, indicates an economic system that engages in thefts, borrowings, and forced removals, violently reorganizing social life as well as the land by thieving resources from Indigenous and Afro-descendent territories. (Gómez-Barris 2017, xvii)

Capturing these acts of theft, brutality and suffering that the global economy of resource production enforces is the intention of this book, illuminating the lives, human and nonhuman, that are ensnared in zones of extraction. Gómez-Barris asks the question: 'in zones of continual extractivism, what responses, engagements, and viewpoints emerge that do not exhaust difference but instead proliferate it?' (Gómez-Barris 2017, xx). For Gómez-Barris, she turns to Indigenous voices whose lands are so routinely stolen and plundered by extractive projects. My own enquiry asks about nonhuman life as another vital witness of capitalist extraction's enactment of violence. Turning to what is elided in the pipelines of commodity production, I thus hope to awaken a consciousness of the nonhuman suffering at the centre of energy production and at its peripheries.

Petrocultures

Petroculture is a growing (sub)discipline in the humanities that focuses on how crude oil has come to shape the cultural and social imaginary of the twenty-first century. Sheena Wilson and Imre Szeman, the co-directors of the research collective, Petrocultures, in the University of Alberta, contend that a transition away from the age of oil requires not just technological and structural transition but also cultural and social change to divest from fossil fuel dependency. Similarly, Szeman and Dominic Boyer argue that,

> The task is nothing less than to reimagine modernity, and in the process to figure ourselves as different kinds of beings than the ones who have built a civilization on the promises, intensities, and fantasies of a particularly dirty, destructive form of energy. (Szeman and Boyer 2017, 14)

The study of petrocultures is therefore a task of not only addressing how we have become ideologically and culturally invested in the fantasy of unlimited energy but to reimagine ourselves as different types of beings, to create new narratives of possibility untethered from destructive and violent forms of energy production. Boyer and Szeman further argue that the critical insights gained from the cultural analysis of energy could begin 'to develop a different relationship to energy' (Szeman and Boyer 2017, 19).

Petrocultures thus diverges from the many historical accounts that focus on the birth and rise of the oil industry. From Daniel Yergin's *The*

Prize: The Epic Quest for Oil, Money, and Power to Peter Maas' *Crude World: The Violent Twilight of Oil*, many of these works are orientated around the politics of a global industry and at times tend to glorify the rise and growth of oil capitalism. There is a sense that while critiquing the corruption of oil companies such as Shell, Chevron and BP, they simultaneously present a triumphant patriarchal capitalism, a world dominated and dictated by oil men. The rise of the fossil fuel economy has been well documented by Andreas Malm's work, as well as Timothy Mitchell's exploration of democracy and carbon-fuelled society. Alternatively, however, humanities scholars such as Barrett and Worden perceive oil not simply as an industry but as a culture: 'a business and a set of aesthetic practices, a natural resource and a trope' (Worden and Barrett 2014, xxi). Such creative engagements with oil in the cultural imaginary allow for utopian thinking—ways of challenging our social and cultural investment in petroleum, and the embedding of oil within our notions of freedom and progress, as well as of the individual and liberty.

In *Living Oil*, LeMenager describes this experience as 'the aesthetics of petroleum' (LeMenager 2014, 68). For LeMenager, there is the question of why, despite environmental destruction and corruption, Americans' love oil. As she notes, 'loving oil has a great deal to do with loving media dependent on fossil fuels or petroleum feedstock from the early to mid-twentieth century, when oil became an expressive form, although often hidden as such, in plain sight' (LeMenager 2014, 66). From classic series such as *Dallas* (1978) (along with its reinvention in 2012) to films such as *Giant* (1956), and more contemporary productions including *There Will Be Blood* (2007), petroleum has captured an important place in the American imaginary. As Huber suggests, 'we often think of our "addiction" to oil as a purely material relation – urban spatial form, disposable plastics – but we need to think more deeply about how our relation to petroleum shapes the way we think and feel about politics' (Huber 2013, 239). Huber's suggestion invites us to understand the ways in which petroleum shapes not only our economy but our social and cultural lives, the way that we think and understand individuality and freedom.

The neoliberal regime that provides the undercurrent for a life dependent on petroleum also feeds into twentieth- and twenty-first-century feminisms. As Sheena Wilson has explored in her work 'Gendering Oil: Tracing Western Petrosexual Relations': 'the histories of feminism and oil are intertwined', with the 'age of oil in the West [...] virtually synchronous with the women's rights movements' (Wilson 2014, 248). For Wilson, the complex relation between the oil industry and the rise of Western feminisms is rife with ironies. The age of oil entails both the advancement of feminism as well as 'reinforcement of long-standing patriarchal conceptualisations of woman as object and as property' (Wilson 2014, 248). Just as animal studies scholar Carol J. Adams describes the sexual and gendered politics of meat, there is a sexual politics to petroleum. Cecily Devereux expands on this gendered dimension of the petroleum industry. Devereux cites Jean Baudrillard who declared: 'All objects, cars included, become women in order to be bought – but this is a function of the cultural system' (Baudrillard and Kendall 2006, 69). This passage, for Devereux, is indicative of a cultural impulse 'to represent both cars as women and women as cars, and thus equally and interchangeably, as commodities' (Devereux 2017, 163). For Devereux, petroculture has informed and shaped the performance of femininity. The complex relation between women's rights movements and the petroleum industry, which acts as both a detriment and force behind Western feminist ideals, demonstrates how entangled the fossil fuel economy is with notions of personhood, rights and freedoms. Yet, as Huber suggests, in order to tackle the impending environmental crisis, 'we need to imagine new ways of life and living that can once again construct popular resonances around notions of public solidarity and the viability of collective management of environmental problems' (Huber 2013, 240). If we are to conceive of a future free from fossil fuel energy, it becomes a task of divesting culturally and collectively from the petroleum industry. By understanding how energy shapes aesthetics and culture, petrocultural analysis presents modes of critique and also possible alternatives.

Whilst petroculture has contended with the gender politics of the fossil fuel industry, it must also look to the colonial and postcolonial histories of oil extraction in the Global South and how these are manifested in cultural narratives and production. As Peter Hitchcock contends, the geoculture of oil should focus not only on what is termed oil's ontology but also oil's 'deep history in the exploitation of the South and of labour' (Hitchcock 2010, 97). Jennifer Wenzel, whose work focuses primarily on Niger Delta, coins the phrase 'Petro-Magic Realism' (Wenzel 2006, 211) to reflect the relation between 'petro-magic' '(a concept from political ecology) with the literary mode of magic realism' (Wenzel 2006, 213). Wenzel focuses on fictions from Ben Okri and Ogaga Ifowodo, examining the ways crude oil seeps into the literary form and highlighting how legacies of imperialism underpin extractive economies and petroculture. Postcolonial theorist

Michael Niblett similarly investigates the cultural role of oil, alongside sugar, and its imbrication in histories of colonial conquest and imperial domination. The oil industry's perpetuation of the legacies of colonialism is further depicted in fictions such as Nawal Saʿdāwī *Love in the Kingdom of Oil* (Saʿdāwī 2019), and Munīf *Cities of Salt* (Munīf 1989). These fictions address crude oil not just as a cultural imaginary and aesthetics but also explore the fossil fuel economy's material and environmental consequences in areas of the Global South. From the American dreams of freedom in narratives like Jack Kerouac's *On the Road* (Kerouac 2011), invested in a world of petroleum, to novels such as Helon Habila's *Oil on Water* (Habila 2011) that explore the devastating effects of the oil industry in the Niger Delta, petrocultures map the cultural manifestations of the petroleum pipeline from extraction to consumption, from oil guzzling regions to peripheralised production zones.

Like Wilson's and Wenzel's important interventions exploring the colonial and gender politics of petroculture, I too propose an intervention which responds to the silence within the field on the effects of the fossil fuel industry on our more-than-human world. As previously noted, I draw on Doherty's suggestion of advancing 'a non-anthropocentric imagining of oil' (Doherty 2014, 367), by foregrounding the materiality of animal bodies and suffering in the context of energy extraction and production. By situating the nonhuman in petrocultural analysis, I aim to address the ways that energy extraction impacts nonhuman life. Boyer and Szeman describe energy humanities as producing a 'speculative impulse' (2017, 22), which offers different ways of perceiving and imagining future worlds. Providing this multifaceted conception of energy and its impact enables a non-anthropocentric mode of thinking, capturing the various ways energy regimes and consumption affects nonhuman life.

For Patricia Yager, literature can be classified according to its relations with resources from wood and whale oil to gasoline to atomic power (Yaeger 2011, 305). Energy resources can shape narrative structures, providing the fuel to transport characters to distant climes, or create the pathways to pivotal events from horsepower in *Wuthering Heights* (Brontë 2003) to gasoline within the road journeys of Kerouac. Yager suggests that 'we need to contemplate literatures' relation to the raucous, invisible, energy-producing atoms that generate world economies and motor our reading' (Yaeger 2011, 307). Similarly, my project here endeavours to capture the resources of fiction and cultural production while incorporating the place of the nonhuman as simultaneously a resource commodity and a victim and witness of extractive capital.

Alliance of Fields: Petrocultures and Animal Studies

Initially, the study of petroculture was a response to Amitav Ghosh's question: where is the Great American Oil Novel? He describes a certain muteness in the cultural imaginary regarding the oil encounter, suggesting that 'the history of oil is a matter of embarrassment verging on the unspeakable' (Ghosh 2002, 29). Amitav Ghosh's claim about the invisibility and unspeakable nature of the oil encounter, however, can be contested, for crude oil is not entirely absent from cultural forms but rather proliferates in different ways beyond its own physicality as a material substance, and can instead produce LeMenager's notion of the aesthetics of petroleum. There is not so much a focus on the oil rig or the extraction site than on the life which petroleum enables. In the animal encounter in literary and cultural production, we can also identify a certain erasure in the sense that animals become objects of study, rendered as artistic and metaphorical symbols, imbued with anthropocentric meaning. Animals, and notions of animality, manifest in cultural forms, yet animals' physical and material presence are notably absent from modern life. In John Berger's essay, 'Why Look at Animals', he claims that 'in the last two centuries, animals have gradually disappeared' (Berger 2009, 21). Berger highlights how animals' previous centrality to human life was 'of course economic and productive' (Berger 2009, 3). Horsepower was used for transportation, for work in mills, and other species for food and clothing; animals had previously existed to a proximity to human life and work. In twentieth- and twenty-first-century capitalism, however, Berger claims, 'Everywhere animals disappear. In zoos they constitute the living monument to their own disappearance' (Berger 2009, 26). In looking at the zoo animal, Berger suggests, the zoo visitor is alone. Can something similar be said of animal presences in literature and culture? Does cultural representation of the nonhuman further isolate us from them and produce a deeper silence? How might this lead to a process where animals are rendered with anthropocentric meaning and artistic function? As a field, the questions animal studies raises and its exploration in literature, like petroculture, can serve to offer distinct insights into our relations and how we perform them. Animal studies in literature has focused not solely on the abstraction of animal as allegories, fables, and metaphors, but their material presences, allowing us to rethink the interspecies encounter. Nonhuman encounters in literature and culture can therefore pose and problematise the ethical dimensions of interspecies relations, and amplify the discussion around species extinction and loss of biodiversity, as well as the impacts of climate change on animal life.

Kathryn Kirkpatrick notes that 'originating in the 1970s with ecofeminist work by Carol J. Adams and others on the interconnected oppressions of women, animals, and the environment, the field of animal studies has become a rich multidisciplinary endeavour transforming knowledge in the academy and generating interest among the general public' (Kirkpatrick and Faragó 2015, 2). The animal turn in literary studies draws on a diverse range of approaches from ecofeminism, posthumanism, continental philosophy, biological and social sciences. The vast array of influences is eclectic and cannot be neatly summarised into a cohesive whole. However, like Wenzel's endeavour in *The Disposition of Nature* to 'consider whether and how the literary can be a part of an environmental praxis: reading for the sake of the planet' (Wenzel 2020, 2), the animal turn in literary studies can be argued to be a process of ethical and political praxis. Reading animals becomes a method for understanding how operations of power and the human/animal binary inform systems of oppression, and why reading and literary analysis can benefit other species and creatures. As Jennifer McDonell notes, 'to think seriously about animals on their own terms is to begin to question the co-construction of the categories of the human and the animal that underpins human exceptionalism' (McDonell 2017, 1473). Reading animals, and their cultural manifestations in literature, as well as certain constructions of the 'human', is a central task in the practice of undoing present power relations that shape our multispecies encounters.

For Anat Pick, cultural forms from cinema to literature can demonstrate not only the cliched expression of the 'human condition' but also 'an expression of something inhuman as well: the permutations of necessity and the materiality that condition and shape human life' (Pick 2011, 5). Pick applies a creaturely prism to culture which moves 'beyond an anthropocentric perspective' (Pick 2011, 5). The task of the creaturely then is to trace what Pick terms 'the logic of flesh' (Pick 2011, 6) across textual and cinematic encounters, a corporeal aesthetic and ethics which does not view species as a primary distinction between different creatures. The question the creaturely poses is how might literature and cultural forms revise and revisit our understanding of other species and our own inhumanity? If, as Pick contends, 'the human-animal distinction' constitutes an arena in which relations of power operate in their exemplary purity (i.e. operate with the fewest moral or material obstacles) (Pick

2011, 1), what can storytelling, fiction and culture offer to disrupt these power relations? I turn to fiction and different cultural sources to not only expose the conditions shaping human animal relations, presenting what Matthew Calarco terms 'the disruptive force in animal suffering' (Calarco 2015, 120), but also to stage new multispecies connections. As previously noted, Baker notes that 'nonhuman animals haunt the peripheries of contemporary fiction' (Baker 2019, 1). I thus explore these peripheries addressing how experiments in fiction and art offer ways of turning to 'cultures [and] contexts that are not exclusively human' (Pick 2011, 5).

The act of questioning human mastery and sovereignty over nonhuman life is a task undertaken by animal studies scholars from Carol J. Adams, Josephine Donavan, Cary Wolfe, Robert McKay, Susan McHugh, Erica Fudge and many more. The literary end of animal studies is the imaginative and empathetic identification with other animals' lives and with the philosophical and ethical questions raised by that engagement. The methodological work of animal literary studies includes deconstructing representations of animals that appropriate the animal as merely literary and mythological figures, as well as critiquing the tendency to observe real animals without attempting to meet their gaze (McDonell 2017, 1474).

These central questions that a literary animal studies raises around an empathetic engagement with nonhuman life, and the field's concern with philosophical and ethical questions, can have important resonances and implications for the study of petroculture. If we can begin to identify with other species and perceive their vulnerability and suffering, how might we also begin to recognise other forms of violence enacted against nonhuman life forms? From the plants and soil that are dying from toxic waste and polluting substances to oil spills that pour out into the ocean, the effects of the fossil fuel economy thus invite ethical questions which are aligned with a multispecies ethics.

Significant amounts of scholarship in animal studies have focused on the affective encounter, the power of empathy, individual emotions and structures of feeling to transform human animal relations. Scholarship such as Lori Gruen's work on entangled empathy with primates or Donna Haraway's work on 'becoming-with' companion species foregrounds a sense of mutuality and interrelatedness with other species. Despite this focus in the field on individual affects, however, there has also been structural analysis exploring the intersections of race, species, and gender with works by Marjorie Spiegel and more recently Claire Jean Kim's *Dangerous*

Crossing: Race, Species and Nature in a Multicultural Age. Unpacking human power relations often exposes the intersections with various other forms of oppression and marginalisation across race, species, gender and class. Nevertheless, what is somewhat absent is an analysis of the material conditions of animal life under capitalism, when nonhuman life becomes what Nicole Shukin describes as 'animal capital' (Shukin 2009, 6). Shukin makes a significant intervention into animal studies with her work, which brings together biopolitical critique and animal life with presenting an exploration into the centrality of animals to regimes of capital. My project similarly intends to emphasise the impacts of the capitalist regime on the nonhuman while also drawing on the affective trajectories that inform much of animal studies as a field. My argument is that a multispecies ethics, informed by affects, emotion and philosophical enquiry, can inform a new perspective within the study of petroculture. Additionally, I argue that the methods of analysis developed in petrocultures that focus on commodity production can enrich and inform animal studies.

The focus on individual affects and the transformative encounter with the nonhuman in animal studies analysis differs somewhat from petroculture, which is invested in materialist approaches. I argue for the convergence of these approaches by presenting a focus on the affective potential of the human/animal encounter while at the same time offering an analysis of extractive regimes and capitalist machinations. My work in this book advances the vital questions of animal studies that involve challenging the centrality of the human, invoking a movement to non-anthropocentricism and asking how this affects our relation to energy.

I follow the continental philosophical trajectory in animal studies, informed by figures such as Val Plumwood, Judith Butler and Anat Pick. I thus move away from the rights-based approaches of Peter Singer and Tom Regan in animal ethics and instead advocate for Pick's *Creaturely Poetics* that presents not an extension of humanism but what she terms humanism's radical retraction, discovering the animality or even vegetative aspects of the human itself (Pick 2011, 6). Like Pick, I explore the place of animals in literature and culture as literary and material presences. Following the recent *Palgrave Handbook of Animals and Literature*, I aim to 'find ways to make sense of them as animals, attentive to their portrayal as an account of their own material or experiential reality' (McHugh et al. 2021, 2). The ways in which literature can restage, discover and reimagine our relation to nonhuman others from different species to energy sources open possible alternatives to the violence of the present. Just as petro-critics emphasise the role of culture in attending to energy change and crisis, the place of the nonhuman in the literary and cultural imaginary can offer a renewed relation to our fellow creatures. As McHugh, Miller and McKay argue, 'in creative, poetic hands, [animal] imagery produces new and insightful ways of understanding human life and the world around it' (McHugh et al. 2021, 3). I explore the convergence of the nonhuman and energy extraction in the literary and cultural imaginary, addressing both the material realities of production whilst exploring the speculative impulse of a more ethical perspective towards our wider ecology.

I ask, 'why are animals absent from so many histories of energy development?' (Haynes 2017, 35) and advance such an inquiry by exploring the place of the nonhuman in its relation to energy production and the violent aftermath of extraction within literary and cultural expression. Haynes goes on to note that the most obvious ways that animals are imbricated within energy regimes is through meat production. As Haynes argues, 'the most intimate and obvious use of animal energy is the oldest: we eat the concentrated energy of animals in muscle energy and fat' (Haynes 2017, 36). Both crude oil and nonhuman bodies are rendered commodities for the purposes of either fuelling our bodies or for transportation. Not only are these industries paralleled through the material process of rendering life and biological matter into utilised commodities, but they are also connected by their environmental impact. In the *Ecological Hoofprint: The Global Burden of Industrial Livestock* (Weis 2013), Tony Weis explores the ways the 'meatification' of diets and the growth of industrial livestock is equivalent and in some cases exceeds the environmental destruction of the fossil fuel industry. By exploring the parallels between animal studies and petroculture, I investigate the abstract and material forms of oppression that operate in the domains of human-animal relations and energy politics.

The Marine Resource: The Origins of Petroculture

When considering a combined focus of animal studies and energy culture, the richest and most fruitful place to begin is perhaps with nineteenth-century whaling and the canonical work of *Moby-Dick*. In Alexis Pauline Gumbs' *Undrowned, Black Feminist Lessons from Marine Life* (Gumbs and Brown 2020), she explores a continuum in which the nineteenth-century imperial efforts of exploiting whale blubber for its commercial utility can be aligned with the present neo-imperial regimes of offshore crude oil extraction. Capturing what it might be like to see through the eyes of a whale, she suggests,

> They saw the boats. Heard those same repurposed boats who came to kill them for their successful blubber, the oil that lit the books and blood of slavery. Even now, it is the commercial pursuits of another form of oil that threatens the bowhead whales whose fat fuelled the capitalist project. Bowhead whales have breathed so much history and outlived it too. (Gumbs and Brown 2020, 47)

The expansive capitalist project of a world energy regime can arguably be said to begin with the whale. While cattle had been used for economic and material purposes 'at home' for millennia, the whale represents the hurtling forth of empire off the shores of Britain and America. Herman Melville's character Ishmael makes a revelatory statement when he declares 'whaling is imperial!' (Melville 1851, 24). Melville's narrative offers an insight into the history of commodity and energy production, tracing the beginnings of a globalised world market. Although whale oil was not primarily used for transportation, as crude oil is, *Moby-Dick* illuminates how whaling is an ideological precursor for fossil fuel extraction and our modern day petrocultures. As Campling and Colas note, 'whaling is an ancient fishery traceable to Neolithic cave paintings in Korea, but [...] under capitalism the hunt for whales changed profoundly, quantitatively and qualitatively' (Campling and Colás 2021, 188). The Western form of capitalist whaling as an energy enterprise is often erased from collective consciousness, viewed as part of a barbaric past, no longer seen in relation to our current modes of energy production. In an attempt to revise this history, I see our current modes of production as beginning with this nonhuman exploitation and suffering.

Far surpassing the British and Dutch, the United States was the world leader in nineteenth-century whaling, establishing an energy regime which lit homes across the country and which cemented the long sea voyages as emblematic of the American spirit of risk, adventure and the conquering of frontiers. As Heidi Scott suggests, 'Whaling was the first American industry to make global economic impacts' (Scott 2014, 4). By understanding the ways commercial nineteenth-century whaling provided the seeds of the globalised project of commerce and extraction, we can expose the imperial legacies informing the contemporary petroleum industry. As Zachary Michael Radford suggests in his analysis of *Moby-Dick*,

> Melville's analysis of America's imperial nature in the early nineteenth century was deeply grounded in the ways in which he perceived that the country was not merely concerned with continental expansion, but rather, in the days of Moby-Dick, angling for empire on a global scale. (Radford 2005, 2)

In particular, by examining whaling as an imperial project, tethered to aims of a globalised market, it is easy to consider the ways the sperm whale is an early iteration of Gómez-Barris' 'extractive zone'. For Wenzel, *Moby-Dick* can be read as a narrative that 'disseminates knowledge about commodities: a text that tells commodity stories – or commodities that tell their own stories – and thereby implicate consumers in the forms of violence that surround them' (Wenzel 2020, 55). For our purposes, it is important to consider *Moby-Dick* as a commodity narrative, magnifying the histories of violence which are imbricated within the production of 'natural' energy resources.

In order to map petrocultures' origins and the development of the resource politics in energy production, *Moby-Dick* provides an exemplary insight into the mechanisms and relations of energy culture, as well as the consequences of life forms being rendered lifeless once conceived as a natural resource. For Graeme Macdonald,

> *Moby-Dick* is arguably among one of the first oil novels. Vested in that notable nineteenth century resource industry—whaling—Melville's novel narrates a megalomaniacal hunt and harvesting of a natural resource all over the world. It stands as prototype representation of a process endemic to the global history of oil extraction and petrochemical commerce. (Macdonald 2012, 7)

Moby-Dick thus exists at a crossroad in our entrepreneurial, extractive history: the last gasp of a dying industry soon to be replaced by petroleum. The novel illuminates the convergence of animal studies and petroculture, as live animals, rather than long-dead ones in the form of crude oil, are at the centre of this nineteenth-century energy regime. As a 'commodity narrative' (Brodhead 1999, 3), *Moby-Dick* explores the growth of capitalist venture which in the narrative is informed by the frontiers of both the economic, colonial and the scientific. In the opening chapter of *Moby-Dick*, Ishmael describes his desire to board a whaling ship as an 'everlasting itch for things remote' (Melville 1851, 5). It is this desire for what is external to the self, what exists outside of the body, and beyond borders of human discovery, which characterises the unrelenting quest of *Moby-Dick*. There is a tension in the narrative between a colonial imperative to master versus a playful curiosity towards the nonhuman world. It can be argued that violent conquest and the colonial venture is characterised by this 'everlasting itch for things remote', the vast expanse of the seas perceived as territory to occupy, manage and control, and most significantly to extract what they discover. In Philip Armstrong's reading of *Moby-Dick*, however, he suggests that under a literary animal studies lens the text invites the question of 'what is at stake ultimately in our ability to think beyond ourselves?' (Armstrong 2005, 93). For Armstrong, the novel's crucial focus is agency, revealing the struggle between the construction of animality, the whale as a material presence and its symbolic attributes. The points and questions raised by Armstrong suggest how Melville's narrative invites a quest into the unknown. The relentless search for the infamous sperm whale could be argued to be a colonial exploration of the undiscovered world with the impending demand to categorise, narrate and conquer. However, in the early stages of the text, there is a focus on capturing the whale for its commercial utility as an energy product or examining it under a scientific gaze. In certain cases, the whale's body becomes a frontier to be crossed, mapped out, and understood through the signifiers of commerce and rational Enlightenment.

In the case of *Moby-Dick*, the marine mammal becomes the resource for profit and consumption. As Armstrong notes, 'Ishmael's descriptions pay less attention to the physiology of the whale of each species and more to its commercial utility' (Armstrong 2005, 111). For the nineteenth-century reader, the whale was mere commodity. Nineteenth-century whaling products were advertised in the spirit of the hunt, a truly violent affair, built upon the rhetoric of the American frontier. Whaling romanticised the act of killing for profit as an epic quest, encountering a sublime spectacle that ultimately led to the whale's demise. Heidi Scott suggests that Melville's narrative can align with the commercial advertisements of nineteenth-century whaling, which depicted a violent excitement and imbued mundane domestic products with the entrepreneurial spirit of the modern American. As Scott further notes, it is an advertisement and tradition that belongs to 'the age of frontier adventure, risk, and danger' (Scott 2014, 12). Whaling's frontier imagery is connected to the cognitive dissonance that shapes modern petroleum consumption in the sense that there is a clear separation between human, animals and nature. Both industries are indeed shaped by what ecofeminist Val Plumwood coins 'hyperseparatism' (Plumwood 1993, 41)—systems of domination and segregation, 'the logic of othering that subjugates not only humans and sentient animals, but also everything else as exploitable' (Plumwood 1993, 5). Through the logic of hyperseparatism, nature and other species become a place of profit. This clear binary logic that segregates the human from their wider ecology shapes the discourse of nineteenth-century whaling, where the sea and its creatures act as a vast frontier. Melville constructs the image of the sea as the whaler's empire, as he declares 'for the sea is his; he owns it' (Melville 1851, 64). At the heart of this imperial project and American expansionism is the interface between the human and nature. Such an opposition can often provide the justification for violence. The whale is thus often mythologised as a 'mortal monster' (Melville 1851, 230), a threat to humanity to be exterminated, where man's 'duty and profit' (Melville 1851, 219) come in direct relation. As John Miller argues, 'the polarity of human and animal is central to imperial mythologies' (Miller 2012, 5). In this imperial discourse, the whale hunt is naturalised as inevitable evolution, a way of assimilating the whale into civilisation through turning it into commodity goods.

Philip Hoare notes how the whale provided a limitless source of production: 'the whale itself was manufactory, of strange substances and of human fortunes' (Hoare 2009, 96). Like our current understanding of fossil fuels, whale oil produced a mythology of wealth and fortune, where each body part from their bones, teeth and blubber became a desired item to feed the demands of industrialisation and capitalist growth. Every 'sailor' became a 'butcher' (Melville 1851, 303), a process described by Melville as the business of 'cutting in', severing the whale's corpse, turning flesh into commodity. Not one element of the whale is left untouched, down to the very jawbone, as witnessed in the narrative when the extraction of the whale's jaw is justified 'for the purpose of extracting the ivory teeth, and furnishing a supply of that hard white whale bone with which fishermen fashion all sorts of curious articles, including canes, umbrella stocks, and handles to riding whips' (Melville 1851, 332). The whale's body is shown as malleable commodity and a variety of domestic goods which no longer resemble the body part from which they were wrought. Such transformations prefigure the use of petroleum chemicals in our contemporary age of plastics. This form of rampant consumerism is poignantly epitomised in Melville's statement: 'Cannibal? Who is not a cannibal' (Melville 1851, 300). It is a process Plumwood would describe as devouring the other who sustains you, assimilating, digesting and incorporating them within the self and into the system of global capitalism.

Moby-Dick at moments, however, forces the reader to acknowledge the life that existed before the animal became a commodity:

> Look at your knife-handle, there, my civilised enlightened gourmand dining off that roast beef, what is the handle made of? – what but the bones of the brother of the very ox you are eating? And what do you pick your teeth with, after devouring that fat goose? With a feather of the same fowl. (Melville 1851, 300)

Melville exposes the cognitive dissonance between consumption and production, pointing to how each commodity item is derived from the animal. It is a literary experiment in making what appears absent present, forcing the reader to acknowledge the origins of their commercial goods. Such detachment also defines the contemporary consumer of petroleum and its derivatives. Just as the nineteenth-century Western citizen was all too familiar with whale-derived goods yet mostly severed from the violent consequences of the whale hunt, so too is our consumption of petroleum detached from the violence of extraction and ecological damage. *Moby-Dick* presents the project of a global capitalism, defining nature as a place for profit and forging the image of the modern consumer, detached and severed from the animal who once existed. Phillip Hoare suggests that *Moby-Dick* is 'a story encoded with its own terrible beauty, one that saw into the future even as it looked into the past' (Hoare 2009, 173). *Moby-Dick's* discourse of commerce, the imagery of the frontier and the scientific gaze anticipate the rhetoric surrounding the discovery of oil in Pennsylvania in 1859. Like whale oil, fossil fuels propagated the story of fortune and wealth and similarly began to shape global commerce, transforming our infrastructure, and the ways we interact with nature.

Nineteenth-century whaling illuminates the animal's role in the course of energy history as an essential part of its beginnings and *Moby-Dick* is a literary rendition of the marine mammal's necessity to the early stages of this global energy regime. This is the basis on which I open my exploration of animal studies and oil culture to move towards more contemporary imaginings of how the nonhuman is part of energy culture and the ways animal exploitation has been inseparable from energy culture history. My ambitions, however, go beyond observing the ways nonhuman animals are a part of energy culture; I go on to address how as fields of study and theoretical practices, the two fields of animal studies and petroculutre are richer and more powerful in tandem. My main claim is thus to suggest that a multispecies ethics and an attentiveness to animal alterity have important significance and ramifications for our energy present and future.

Chapter Outlines

This book centres the relationship between nonhuman exploitation and energy extraction, exploring the literary and cultural representations of the nonhuman's position in the violent histories of energy regimes. The next chapter explores the theoretical foundations to address such an area of my research. My contention is that a politics of vulnerability is an essential antidote to and disruptive force against extractive regimes. This theoretical chapter thus addresses the role of vision and operations of displacement which I've briefly addressed here in the introduction and moves then to explore the role of bodily and earthly vulnerability, and the ways its power offers an environmental and multispecies ethics. To discover the suffering that is often erased and hidden, I turn to Carol J. Adams' concept of the 'absent referent' alongside postcolonial scholar Rob Nixon's notion of 'slow violence'. The intention of utilising these conceptual tools together is to illuminate an environmental and animal politics which where each informs and empowers the other. These conceptual tools of rendering visible what is often not apparent move us towards an ethics and politics of vulnerability. My ideas of bodily and earthly vulnerability allow me to implicate a non-anthropocentric politics in the domain of energy culture. Throughout later chapters, my readings of literary and cultural works will be informed by this underpinning of the disruptive and illuminating force of a corporeal yet planetary vulnerability. I look to what happens when we discover this vulnerability in the space of the extractive zone and if its recognition has the influence and power to create change.

The third chapter addresses the operations of displacements and parallels between industrial animal slaughter and crude oil extraction through two main literary works, Michel Faber's *Under the Skin* and Nnedi Okorofar's *Lagoon*. Here, I open to consider the implicit anthropocentrism and colonial manifestations of the 'human' as a way of understanding how these industries are informed by a liberal humanism, which permits life to be killed or suffer in the name of capitalist expansion and development. Drawing on Kathryn Yusoff's insights that 'colonialism was an extraction project' (Yusoff 2018, 89), the chapter examines the systematic erasures of meat and oil industries while exploring how they are ideologically informed by colonial and neo-colonial power, a power that is tied to a particular conception of the 'human'. My analysis of these literary texts is focused on how they offer a telescoping function of magnifying peripheral scenes, of making what is often absent present—the scene of the slaughterhouse and the site of crude oil production. Towards the latter stages, I trace moments of disruptive vulnerability in the texts, where there are moments of resistance through the encounter with an earthly and planetary vulnerability.

From examining modes of production to moving towards their consequences, I move to consider the impacts of oil spills, contaminated water ways and roadkill. Here, I explicitly address the infrastructures of energy and automobility which impact nonhuman life. The three cultural sources I explore are Steve Baker's photography of roadkill, the oil spills and environmental devastation in Helon Habila's *Oil on Water* (Habila 2011), and the graphic novel and poetry of Rita Wong's *Perpetual* (Wong 2015). The opening of the chapter introduces the role of affects when one is witness to environmental ruin and nonhuman death. Drawing on Judith Butler's analysis of grievability, Sara Ahmed's cultural politics of emotion, and Sianne Ngai's exploration of disgust, I look to how a role of affects come to the fore in the exposure of vulnerability. Through an analysis of these cultural works that explore hope in the space of the extractive zone and glimmers of resistance in scenes of environmental devastation, I examine the moments of rupture and recognition which emerge through exposure to vulnerability. I thus suggest how these unruly affective registers offer a path to a transition from our fossil fuel dominated economy.

The subject of energy transition and the paths to arriving there are at the centre of the final chapter. My contention is that introducing cultural works which allow for a non-anthropocentric perspective in the domain of energy culture can inspire new ways of imagining our energy-intensive lifestyles. In Chap. 5, I explore three different science fictional works: Cixin Liu's 'Moonlight' (Liu 2019), Ursula K. Le Guin's *The Dispossessed* (Le Guin 2014) and Stanislaw Lem's *Solaris* (Lem 2003). Opening with how science fiction is an exemplary genre to contend with both animal alterity and speculative energy resources, I emphasise how such a genre opens us to alternative paths from our present. Liu's short story 'Moonlight' offers a precautionary tale to the redundant path of green capitalist solutions to the energy crisis, while I suggest *The Dispossessed* and *Solaris* offer a way of unthinking the mastery of the human and capitalist imperial expansion. Drawing on the dehumanist and decolonial approach of Julietta Singh's *Unthinking Mastery* (Singh 2018), I hope to illuminate the ways these cultural works offer a mode of unlearning and unthinking which enables us to arrive at a new way of approaching our environmental and energy futures. This book thus offers new terrain to think through our relation to energy and our nonhuman companions, discovering the ways environmental thinking and animal justice should be considered as a combined fight for resistance and change. The politics of vulnerability I advocate for emphasises an earthly and planetary sense of belonging and fellowship with other creatures and the world around us. It is a plea for change and ethical attention to those often erased from our moral and political considerations. Weaving feminist and decolonial approaches, I illuminate tools of recognition through the power of vulnerability which goes beyond victimisation and shows how being vulnerable, recognising vulnerability within others as well as within ourselves, and the shared experience of it, offers a powerful antidote to a world based on accumulative and relentless extraction.

References

After Oil. 2016. Edmonton, Alberta: Petrocultures Research Group.

Armstrong, Philip. 2005. What Animals Mean, in *Moby-Dick*, for Example. *Textual Practice* 19 (1): 93–111. https://doi.org/10.1080/0950236042000329663.

Baker, Timothy C. 2019. *Writing Animals: Language, Suffering, and Animality in Twenty-First-Century Fiction.* Palgrave Studies in Animals and Literature Ser. Cham: Palgrave Macmillan US.

Baudrillard, Jean, and Stuart Kendall. 2006. *Utopia Deferred: Writings from Utopie (1967–1978).* New York: Semiotext(e).

Berger, John. 2009. *Why Look at Animals?* Great Ideas 80. London New York Toronto: Penguin Books.

Brodhead, Richard H., ed. 1999. *New Essays on Moby-Dick.* Digital repr. The American Novel. Cambridge: Cambridge Univ.Pr.

Brontë, Emily. 2003. *Wuthering Heights.* Edited by Pauline Nestor. Reiss. with revisions and a New preface. Penguin Classics. London: Penguin Books.

Calarco, Matthew. 2015. *Thinking Through Animals: Identity, Difference, Indistinction.* Stanford University Press.

Campling, Liam, and Alejandro Colás. 2021. *Capitalism and the Sea: The Maritime Factor in the Making of the Modern World.* London; New York: Verso.

Devereux, Cecily. 2017. 'Made for Mankind': Cars, Cosmetics, and the Petrocultural Feminine. In *Petrocultures, Oil, Politics, and Culture*. McGill-Queen's Press.
Doherty, Melanie. 2014. Oil and Dust, Theorising Reza Negarestani's Cyclonopedia. In *Oil Culture*. University of Minnesota Press.
Federici, Silvia. 2020. *Beyond the Periphery of the Skin: Rethinking, Remaking, and Reclaiming the Body in Contemporary Capitalism*. Kairos. Oakland: PM Press.
Fowkes, Maja, and Reuben Fowkes. 2022. *Art and Climate Change*. World of Art. London; New York: Thames & Hudson.
Ghosh, Amitav. 2002. Petrofiction: The Oil Encounter and the Novel. *The Imam and the Indian* 75.
Ghosh, Amitav. 2021. *The Nutmeg's Curse: Parables for a Planet in Crisis*. London: John Murray.
Gómez-Barris, Macarena. 2017. *The Extractive Zone: Social Ecologies and Decolonial Perspectives*. Dissident Acts. Durham. London: Duke University Press.
Gumbs, Alexis Pauline, and Adrienne Maree Brown. 2020. *Undrowned: Black Feminist Lessons from Marine Mammals*. Emergent Strategy Series, No. 2. Chico, CA Edinburgh: AK Press.
Habila, Helon. 2011. *Oil on Water*. London: Penguin Books.
Haraway, Donna. 2016. *Staying with the Trouble: Making Kin in the Chthulucene*. Experimental Futures. Technological Lives, Scientific Arts, Anthropological Voices. Durham London: Duke University Press.
Harootunian, Harry. 2000. *History's Disquiet: Modernity, Cultural Practice, and the Question of Everyday Life*. Columbia University Press.
Haynes, Melissa. 2017. Animals and Energy. In *Fueling Culture: 101 Words for Energy and Environment*. Fordham Press.
Hitchcock, Peter. 2010. Oil in an American Imaginary. *New Formations* 69 (69): 81–97. https://doi.org/10.3898/NEWF.69.04.2010.
Hoare, Philip. 2009. *Leviathan, or, The Whale*. London: Fourth Estate.
Huber, Matthew T. 2013. *Lifeblood: Oil, Freedom, and the Forces of Capital*. A Quadrant Book. Minneapolis: University of Minnesota Press.
Kerouac, Jack. 2011. *On the Road*. London: Penguin Books.
Kirkpatrick, Kathryn, and Borbála Faragó. 2015. *Animals in Irish Literature and Culture*. 1st ed. Palgrave Studies in Animals and Literature. London: Palgrave Macmillan UK. https://doi.org/10.1057/9781137434807.
Klein, Naomi. 2013. Dancing the World into Being: A Conversation with Idle No More's Leanne Simpson. *Yes! Solutions Journalism*. https://www.yesmagazine.org/social-justice/2013/03/06/dancing-the-world-into-being-a-conversation-with-idle-no-more-leanne-simpson.
Le Guin, Ursula K. 2014. *The Dispossessed*. First Perennial Classic Paperback edition, Reissued in Harper Perennial Modern Classics. Harper Perennial Modern

Classics. New York, NY London Toronto Sydney New Delhi Auckland: Harper Perennial.

LeMenager, Stephanie. 2014. *Living Oil: Petroleum Culture in the American Century*. Oxford Studies in American Literary History 5. Oxford; New York; Auckland; Cape Town; Dar es Salaam Hong Kong Karachi; Kuala Lumpur; Madrid; Melbourne; Mexiko City: Oxford University Press.

Lem, Stanislaw. 2003. *Solaris*. Faber & Faber.

Liu, Cixin. 2019. "Moonlight." In *Broken Stars*. Head of Zeus.

Macdonald, Graeme. 2012. Oil and World Literature. *American Book Review* 33 (3): 7–31. https://doi.org/10.1353/abr.2012.0079.

McDonell, Jennifer. 2017. The Animal Turn, Literary Studies, and the Academy. In *Literary Theory: An Anthology*. Blackwell.

McHugh, Susan, Robert McKay, and John Miller. 2021. *The Palgrave Handbook of Animals and Literature*. Palgrave Studies in Animals and Literature. Cham: Palgrave Macmillan.

Melville, Herman. 1851. *Moby-Dick, or, The Whale*. Penguin Classics. New York: Penguin Books.

Miller, John. 2012. *Empire and the Animal Body: Violence, Identity and Ecology in Victorian Adventure Fiction*. Anthem Press.

Moretti, Franco. 2000. Conjectures on World Literature. *New Left Review*, January. https://newleftreview.org/issues/ii1/articles/franco-moretti-conjectures-on-world-literature.

Munīf, ʿAbd al-Raḥmán. 1989. *Cities of Salt*. New York: Vintage international.

Pick, Anat. 2011. *Creaturely Poetics: Animality and Vulnerability in Literature and Film*. New York: Columbia University Press.

Plumwood, Val. 1993. *Feminism and the Mastery of Nature*. Opening Out. London: Routledge.

Radford, Michael. 2005. The Whale and the World in Melville's Moby-Dick: Early American Empire and Globalization. University of Montana. scholarworks.umt.edu/etd/1108.

Saʿdāwī, Nawāl. 2019. *Love in the Kingdom of Oil*. Translated by B. Hatim and Malcolm Williams. London: Saqi Books.

Sands, Danielle. 2019. *Animal Writing: Storytelling, Selfhood and the Limits of Empathy*. Edinburgh University Press. https://doi.org/10.1515/9781474439053.

Scott, Heidi. 2014. Whale Oil, Consumerism, and Modern Conservation. In *Oil Culture*. University of Minnesota Press.

Shukin, Nicole. 2009. *Animal Capital: Rendering Life in Biopolitical Times*. Posthumanities 6. Minneapolis: University of Minnesota Press.

Singh, Julietta. 2018. *Unthinking Mastery: Dehumanism and Decolonial Entanglements*. Durham London: Duke University Press.

Szeman, Imre, and Dominic Boyer, eds. 2017. *Energy Humanities: An Anthology*. Baltimore: Johns Hopkins University Press.

Warwick Research Collective. 2015. *Combined and Uneven Development: Towards a New Theory of World-Literature*. Postcolonialism across the Disciplines 17. Liverpool: Liverpool University Press.

Weis, Anthony John. 2013. *The Ecological Hoofprint: The Global Burden of Industrial Livestock*. London, England: Zed Books. https://doi.org/10.5040/9781350223080.

Wenzel, Jennifer. 2006. Petro-Magic-Realism: Toward a Political Ecology of Nigerian Literature. *Postcolonial Studies* 9 (4): 449–64. https://doi.org/10.1080/13688790600993263.

Wenzel, Jennifer. 2020. *The Disposition of Nature: Environmental Crisis and World Literature*. 1st ed. New York: Fordham University Press.

Wilson, Sheena. 2014. Gendering Oil Tracing Western Petrosexual Relations. In *Oil Culture*. University of Minnesota Press.

Wilson, Sheena, and Andrew Pendakis. 2012. Sight, Site, Cite: Oil in the Field of Vision. *Imaginations: Journal of Cross-Cultural Image Studies* 3 (2): 4–5. https://doi.org/10.17742/IMAGE.sightoil.3-2.1.

Wong, Rita. 2015. *Perpetual*. La Vergne: Harbour Publishing Co. Ltd.

Worden, Daniel, and Ross Barrett, eds. 2014. *Oil Culture*. Minneapolis, MN: University of Minnesota Press.

Yaeger, Patricia. 2011. Editor's Column: Literature in the Ages of Wood, Tallow, Coal, Whale Oil, Gasoline, Atomic Power, and Other Energy Sources. *PMLA/Publications of the Modern Language Association of America* 126 (2): 305–326. https://doi.org/10.1632/pmla.2011.126.2.305.

Yusoff, Kathryn. 2018. *A Billion Black Anthropocenes or None*. Forerunners Ideas First from the University of Minnesota Press. Minneapolis, MN: University of Minnesota Press.

CHAPTER 2

Theoretical Foundations: Politics of Attention and an Ethics of Vulnerability

Simone Weil's philosophy centres upon the unravelling of the self through severe affliction, the process of losing the ability and agency to say 'I'. The experience of affliction for Weil is fundamentally tied to bodily vulnerability. Weil contends that 'the vulnerability of precious things is beautiful because vulnerability is a marker of existence' (Weil 2002, 108). Like Julietta Singh, Weil understands vulnerability as introducing not only the risk of destruction but presenting something transformative, revealing other ways of being in and for the world, as well as beyond it (towards the divine). Julietta Singh claims:

> The practice of vulnerable reading can move us "beyond" mastery, not in the sense of exceeding it but in the sense of surviving it in order to envision being otherwise in and for the world. (Singh 2018, 23)

The practice of vulnerable reading suggests that vulnerability is not entirely negative, as she offers a depiction of vulnerability which presents ways of becoming together in common. Weil's concept of affliction is the inevitable result of bodies and lives that are and remain vulnerable and dependent on what exists outside of themselves. To be vulnerable is to be subject to pain and suffering, open to brute and external force; in Weilian terms, it is to be exposed to the force of gravity—which is the weight of

J. Taylor, *The Nonhuman Narratives of Energy*, Palgrave Studies in Animals and Literature,
https://doi.org/10.1007/978-3-031-94332-4_2

the material world. Vulnerability and its various implications, from affliction and pain, to mutual co-dependency and fellowship, can however offer sites of resistance and hope in moments of despair and violence. Weil's statement that vulnerability is a marker of existence provides the focal point of Anat Pick's *Creaturely Poetics*, which foregrounds a non-anthropocentric ethics and aesthetics, discovering fellowship through vulnerable co-existence across species lines. If vulnerability can serve as a disruptive antidote to the violence against nonhuman others, how might such ruptures foster a challenge to capitalist machinations and the commodification of life? Whilst vulnerability may leave us, in Judith Butler's words, radically 'undone' (Butler 2006, 23), the purpose of this chapter is to demonstrate that vulnerability is a productive modality and site of resistance, offering a position of empowerment over victimhood and despair.

Informed by theorists of vulnerability I offer a frame to understand planetary and earthly fellowship through a commonality found in shared vulnerability, which is not limited to being embodied but is connected to a sense of whether one or something flourishes or perishes, an earthly connection to being subject to force and external power and at times offering a resistance to that force. This earthly vulnerability I speak of is inherently beyond anthropocentrism as it does not privilege sentience or subjectivity for whether life or matter is of moral value and political worth. Rather than a posthumanist formulation, it is more similar to Julietta Singh's conception of dehumanism in that I explicitly invoke conversations with the decolonial and foreground vulnerability as a compass to unpack human exceptionalism, imperial legacies and liberal capitalism. I refer to a planetary and earthly vulnerability as an invitation to challenge the suffering and devastation wrought on nonhuman creatures and the earth itself.

The politics of visibility which I suggest shapes the ambitions and challenges of petroculture and animal studies will be explored through introducing Rob Nixon's 'slow violence' and Carol J. Adams 'the absent referent'. I argue that the concept of 'slow violence' (Nixon 2013, 1) advanced by Nixon and Adams' notion of the 'absent referent' (Adams 2020, 40) serve as tools of exposure and recognition. In Nixon's words, they point towards the violence that is not typically considered violence at all (Nixon 2013, 2). These methods of exposure focus on the challenges of representation, the cognitive dissonance that often shapes consumption and production, and the systematic and material erasures which occur through commodification.

I challenge such erasures and violence through a focus on vulnerability as advanced by three crucial thinkers: Judith Butler, Simone Weil and Anat Pick. The radical formations of vulnerability, dependency, grief, attention and fellowship are concepts which emerge from these theorists' explorations of violence and its possible antidotes. Using vulnerability as a possible steppingstone to developing just and feminist environmental and energy futures, I intend to explore each theorists' understanding of vulnerability as a mode of resistance to extractive economies, thus forming the core principles of what I term a planetary and earthly vulnerability.

Important to my readings is the neo-colonial context of extraction, I thus also draw on the scholarship of Kathryn Yusoff, Sylvia Wynter, Julietta Singh and Alexander G. Weheliye. I explore the intimacies of the colonial project of extraction alongside humanist discourse and ideology. If, as Pick suggests, the 'human is a tenuous, fragile construct' (Pick 2011, 27), it is also an ideological one invested in legacies of imperialism that draw oppressive lines of distinction, delineating the 'human' from its supposed 'others'.

My approach of building alliances across often separated and distinct schools of thought offers a way of understanding and revealing the space and violences of the extractive zone, revealing the hegemonic structures which shape modes of production and obscure their consequences, whilst showing how the present violence of petroculture might be resisted and exploring potential alternatives. I look to the discovery of a planetary and earthly vulnerability as both a unifying force across discourses and a possible rupture in practices of violent extraction. It is through building paths of solidarity across difference that we 'might begin to construct the provisional infrastructure of a new world amidst the ruins of the old' (Out of the Woods Collective 2020, 5). As the Out of the Woods Collective go on to suggest, 'the ecological crisis is a product of centuries of this system, of innumerable extractions and exploitations, indescribable enslavements and extirpations' (2020, 4). Imre Szeman and Jennifer Wenzel note the multifaceted aspects of extraction describing 'the relationship between resource extraction as a moment and process under capitalism (or socialism), and extractivism as an ideology and cultural logic that permeates social imaginaries as well as literary and other discourse' (Szeman and Wenzel 2021, 1). Through an ethics of vulnerability, I forge an anti-extractive paradigm in response to violent practices of extraction as they are articulated and visualised through literature and culture.

Methods and Tools of Exposure

Carol J. Adams' ground-breaking work in *The Sexual Politics of Meat* offered what she termed as feminist-vegetarian critical theory. The concept I wish to explore in greater detail is that of the 'absent referent' (Adams 2020, 20). Nixon's *Slow Violence and Environmentalism of the Poor* is a similarly pivotal work which articulates what can be the slow-moving atrocity of ecological damage through his concept of 'slow violence' (Nixon 2013, 1). Both scholars have different ambitions: Adams focuses mainly on animal and feminist ethics and meat consumption, whereas Nixon aims to articulate the devastation of the environment through literature in areas of the Global South. Nevertheless, there are commonalities shared between these two theorists involving representation, recognition and exposure. I use these concepts in tandem to point towards the nonhuman suffering and ecological destruction which can often be erased from cultural focus.

Adams' absent referent articulates the severance and cognitive dissonance at the heart of meat eating. When purchasing meat we do not associate the neatly packaged and processed item or commodity with the living, breathing animal it once was. The absent referent thus functions as a form of displacement, detaching us from any ethical recognition which might hinder our consumptive habits. In Adams' account, this displacement functions in three ways. Firstly, there is a physical displacement, in which animals are slaughtered far from the view of the consumer, sequestered in an enclosed location from which their suffering can be neither seen nor heard by the public which will consume them. Secondly, there is a linguistic displacement, in the sense that we adopt a different style to address the animal we are about to consume: they are spoken of in the aesthetic terms of a cuisine rather than through terminology which might suggest a formerly living creature. The third operation is that of metaphor, where the actual figure of the exterminated animal is replaced with a metaphorical use of meat as a reflection of one's experience in terms of an inert product: to 'feel like a piece of meat' is not to recognise animal meat for what it once was, but to think of oneself in terms of the manipulated object we have made. These three functions produce a selective amnesia which enables consumption without any ethical obligations towards the nonhuman animal.

For Adams, 'the absent referent can be anything whose original meaning is undercut and absorbed into a different hierarchy of meaning' (Adams 2020, 21) and for our purposes I will be reapplying the concept to petroculture. To that end the most significant dimension will be the physical displacement of the effects of oil consumption. Like the meat industry, the effects of oil culture are shaped by a sense of displacement, whereby we enjoy the pleasures permitted by our fossil fuel lifestyles with little or no regard for the consequences of environmental degradation. Such a parallel has been noted by Melissa Haynes in her analysis of the animal's role within energy development:

> Just as oil extraction sites stand at distance from gas stations, concentrated animal feeding operations and meat packaging plants are kept at a remove from urban centres, and bloodless packages of beef and pork are hard to imagine as ever having been cows or pigs. (Haynes 2017, 36)

I advance Haynes' argument by elaborating on the parallel mechanisms which shape both energy production and the treatment of nonhuman animals. The combination of the absent referent and slow violence allows me to address and magnify these operations of displacement and abstraction.

Nixon opens *Slow Violence and Environmentalism of the Poor* with an epigraph from Arundhati Roy:

> I think of globalization like a light which shines brighter and brighter on a few people and the rest are in darkness, wiped out. They simply can't be seen. Once you get used to not seeing that something, then, slowly, it is no longer possible to see it. (Roy et al. 2004, 1)

Nixon's choice of epigraph indicates the uneven nature of representation in our globalised world, within which our focus is directed only to certain lives, tragedies and events by the mass media. Like Adams' absent referent, Nixon hopes to overcome these representational challenges by using theoretical tools of exposure. Nixon conceives of the concept of slow violence as a way of capturing the events elided from view due to the 'spectacle deficiency', which ensures they are not considered newsworthy. He defines slow violence as 'a violence that occurs gradually and out of sight, a violence of delayed destruction that is dispersed across time and space, an attritional violence that is typically not viewed as violence at all' (Nixon 2013, 2). What unites Adams' and Nixon's work is their focus upon what 'remains outside our flickering attention spans- and outside the purview of a spectacle-driven corporate media' (Nixon 2013, 6). Slow violence and the absent referent can thus both help us to draw attention to the often invisible violence of petroculture. Despite Nixon's specific focus on environmental damage and Adams' focus on the suffering of the nonhuman animal, the central concern for both theorists is that of recognition. Connecting these two concepts will therefore serve as the ground for a radical work of exposure, which reveals the connections between a planetary crisis and the continued oppression of the nonhuman animal. Nixon and Adams point us towards this crisis in representation, whereby they acknowledge the environmental damage and animal suffering which is absent from consciousness. Narrative and art forms from the periphery therefore offers ways of contending with this crisis in representation. For Nixon and Adams, it is a case of 'making the absent referent present' (Adams 2020, 22), producing a 'different kind of witnessing: of sights unseen' (Nixon 2013, 15). The literature and culture I examine expose these different kinds of witnessing, producing presence in the space of absence.

But how might we cultivate such recognition, allowing the effects of our fossil fuel culture to be revealed? For this I turn to three exemplary thinkers of exposure and vulnerability in the fields of feminism, philosophy and animal studies: Judith Butler's work on grief and vulnerability, Simone Weil's philosophy and decreative ethics, and Anat Pick's *Creaturely Poetics.* Weil's extraordinary statement that 'the vulnerability of precious things is beautiful because vulnerability is a mark of existence' (Weil 2002, 108), encapsulates what I wish to draw from these figures: an ethics and politics which incorporates a responsibility towards the animal Other grounded not in their possessing particular (human) qualities but instead from a recognition of their vulnerability. Having vulnerability as our focal point allows us to turn to a 'different kind of witnessing', which produces 'sights of the unseen' (Nixon 2013, 15), addressing all life and thereby including all that which falls outside of our current ethical radar. Attuning ourselves to the vulnerability of other species and the earth itself can lead us to an ecological awareness and ethics, which would enable us to escape our fossil fuel intensive lifestyles and engage in more sustainable practices and envision feminist and just energy futures.

Radical Earthly Vulnerability: Butler, Weil and Pick

Val Plumwood's account of becoming prey serves as insightful introduction to the powers of becoming vulnerable and what ramifications it may hold. Plumwood details a personal account of 'becoming prey' in the event of an attack by a crocodile. In response to the media's exaggeration and demonisation of her attacker, Plumwood writes about the concept of invulnerability—something she views as 'typical of the mind of coloniser' (Plumwood 1995, 34). The experience of being prey allowed for a reflection on the demarcations the West builds between itself and nature. As Plumwood notes, 'in the West, the human is set apart from nature as radically Other' (Plumwood 1995, 34). For Plumwood, the outrage invoked by the media over the incident of her becoming prey is nothing but a reminder of our own animality, in which we are also vulnerable to others and the terrifying prospect of becoming meat. The implication of such an attack is the unbearable reality of our own radical vulnerability.

Butler, Weil and Pick's philosophical thought and analysis is anchored in such conditions of vulnerability. Like Plumwood, however, the revelation of vulnerability leads not to a cry of despair but to the building of solidarity, of recognising a multispecies and an earthly community. In what follows, I detail central concepts from Butler's later work on grief, Weil's philosophical enquiries into attention and decreation, and Pick's creaturely fellowship and notion of 'letting be'. The operations of exposure and recognition remain central modalities of the following exposition, which aims to demonstrate vulnerability as a force of collective resistance. The purpose of bringing these three thinkers together is to explore how vulnerability and affect leads to a path towards fellowship and an earthly solidarity. What I explore through these three thinkers of vulnerability can be illuminated in Yoon Sook Cha's reading of Weil in which she suggests there is a 'non-cognitive modality' (Sook Cha 2017, 3) at work when hearing the cry of the suffering. This 'non-cognitive modality' is a way of introducing a non-anthropocentric ethics and politics that I suggest vulnerability has the potential to release and empower.

Butler's work in *Precarious Life*, *Frames of War*, and their more recent book *Force of Non-Violence*, centres on the politics of mourning and bodily precarity. In the earliest account of these concerns, Butler discusses the absence of mourning for the violence that the United States inflicts on regions in the Middle East. As Butler argues,

> There are no obituaries for the war casualties the United States inflicts, and there cannot be. If there were to be an obituary there would have had to have been a life, a life worth noting, a life worth valuing and preserving, a life that qualifies for recognition. (Butler 2006, 34)

Butler describes the uneven distribution of grievability, outlining a certain politics and hegemonic structure to acts of mourning. While Butler's work focuses primarily on the human, I wish to extend their insights to the nonhuman animal and the environment by acknowledging the ways we are embedded in more-than-human worlds. Their work invites the question of how we are to form an ethical response to those we cannot see, whose names we cannot know, to those whose form and face are not familiar and even those who do not possess a face at all. How might such questions allow us to think ecologically with Butler's work, moving beyond the domain of the human to consider the vulnerability of all life forms?

For Butler,

> An obligation does emerge from the fact that we are, as it were social beings from the start, dependent on what is outside ourselves, on others, on institutions, and on sustained and sustainable environments, and so are, in this sense, precarious. (Butler 2016, 23)

I would argue that Butler's thinking enables us to consider how we are not only mutually dependent on other humans but also on the environment and different species. Becoming aware of our own sense of vulnerability and of the world around us invites a non-anthropocentric perspective, which rejects species hierarchy and human exceptionalism, and allows us to view nature not as entirely instrumental to the human species but instead as comprised of vulnerable agents entitled to flourish.

For Butler, certain states of vulnerability can be induced through the experience of grief. Butler remarks that loss 'makes a tenuous we of us all' (Butler 2006, 20) in that 'something about who we are is revealed, something that delineates the ties we have to others' and which 'shows us that these ties constitute who we are, ties or bonds that compose us' (Butler 2006, 22). Grief evokes a sense of fragility which reveals what makes us who we are and what binds us to other creatures and the world around us, implicating us in lives which are not our own. Butler argues, however, that whoever becomes the subject of mourning is dependent upon our notion of who is recognisably human. The solution for Butler is to focus upon the human face as a buffer against the vicious brutality which derealises certain lives—working against dehumanisation through the recognition of a common humanity and vulnerability which serves to address the anonymous 'Others' excluded from our ethical sphere. Butler frames their argument in the aftermath of September 9/11, a traumatic event which produced a sudden realisation amongst the American people and the West that even they were vulnerable subjects. Reflecting on the implications of injury, grief and vulnerability, Butler suggests,

> To be injured means that one has the chance to reflect upon injury, to find out mechanisms of its distribution, to find out who else suffers from permeable borders, unexpected violence, dispossession, and fear, and in what ways. (Butler 2006, XII)

Our exposure to injury and loss reveals how we are subject to what exists outside of the self. America's response to this vulnerability, however, was the waging of war. The inclusive power of recognising a shared state of precarity and vulnerability, the ability to form a collective 'we', failed in America's case and instead formulated an 'us' and 'them' divide, a shoring up and solidifying of borders, and the proliferation of violence. Butler, however, suggests an alternative non-violent ethics to the exposure to vulnerability and grief, one which can evoke a sense of compassion and solidarity, and inspire a quest for global justice. The mass outrage and grief surrounding 9/11 led Butler to a consideration of how different tragedies and deaths hold an uneven political power. Turning to the prisoners of Guantanamo Bay and those lives lost due to US military invasion, Butler notes how certain lives will be highly protected while others will find no such value or support. They appear as 'faceless' or 'whose faces are presented to us as so many symbols of evil, authorising us to become senseless before those lives we have eradicated, and whose grievability is indefinitely postponed' (Butler 2006, xvii). In the face of such violent exclusion, Butler's argument, however, in this earlier work, remains anthropocentric focused primarily on human events and tragedies.

While I agree with Butler's call for a radical sense of openness towards the Other, I believe this cannot be accessed through an expansion of humanism, but rather through its retraction. Initially, Butler situates their ethics as a response to specifically human tragedies, forgetting that our interactions are built upon more-than-human worlds. This anthropocentrism is revised in their more recent work. As Butler argues,

> An ethics and politics of nonviolence would have to account for this way that selves are implicated in each other's lives, bound by a set of relations that can be as destructive as they can be sustaining The relations that bind and define extend beyond the dyadic human encounter, which pertains not only to human relations, but all living and inter-constitutive relations. (Butler 2020, 9)

As Butler's recent work suggests, which lives we consider grievable has important implications for our relations to other species and the environment itself. If grief and vulnerability possess a transformative power, igniting a sense of solidarity and interdependence, they can also serve to create more just ways of living and a way to empathetically engage with the world.

In order to mourn a lost life, however, it must be considered a life in the first place. As Butler notes, 'there are subjects who are not quite recognisable as subjects, and there are lives that are not quite – or, indeed, are never - recognised as lives' (Butler 2016, 4). It is my intention to not focus primarily on subjectivity as the basis of our ethics and grounds for the work of mourning. Instead, in order to mourn beyond the human, we must utilise Butler's notion of vulnerability as a way of recognising nonhuman life.

Mourning can perhaps be a tool to respond to ecological and planetary crisis. As Ashlee Cunsolo notes, 'we need the mechanisms to extend grievability to non-human bodies and recognise them as mournable subjects, particularly within the discourses of climate change' (Cunsolo 2017, 141). The political implications of who is to be considered grievable and who is not provide an insight into the operations of power that determine which bodies are publicly valued over others. To highlight the deaths left outside of public obituaries and the bodies denied social intelligibility, to mourn those deemed unmournable can become a powerful form of resistance.

Cunsolo cites evidence of the transformative and affective power of mourning during the 1990s AIDS crisis protests. She does not try to equate the lives of nonhuman animals and the environment with those lost to the AIDS epidemic but instead aims to highlight how the activists' struggle for recognition and how they channelled grief and loss into political action. Mourning in this way presented a form of resistance which served to empower bodies that had been denied ethical value. For Cunsolo, this type of mourning is a recognition of the lives of the oppressed and serves to enforce solidarity and inspires intervention, providing a framework we can apply to climate change, utilising the grief we feel over the loss of biodiversity, nonhuman suffering and environmental destruction to create political and environmental change. In Deborah Gould's research on the AIDS activist group ACT UP, she notes: 'affect, in short, has the potential to escape social control, and that quality creates greater space for counter hegemonic possibilities' (Gould 2009, 39). Gould's account of the period notes how death and loss surrounded the activists and how 'mourning became militancy within the movement' as grief turned to anger and action (Gould 2009, 8). Their public displays of mourning and political funerals served to 'redefine the AIDS body as something mournable and something absolutely imperative to grieve publicly and openly' (Cunsolo 2017, 147). It became a way of reconstituting the rights of subjects who were denied social intelligibility and of demanding the recognition of those lost to the epidemic within public discourse. As Eve Sedgewick states, 'it's been one of the great ideological triumphs of AIDS activism that, for a whole series of overlapping communities, any person living with AIDS is now visible' (Sedgwick 1993, 24). The movement utilised the power of affect, mourning and the art of noticing to demand change in how the victims were treated by government bodies and wider society.

Although I am not attempting to conflate the AIDS activists' struggle for health care, structural change and recognition with animal ethics and environmental protest, the revelation of how marginalised bodies became ethically and politically relevant through the affective and transformative effects of mourning has parallel possibilities for igniting environmental and animal justice. Memorialising death and destruction—human, animal and environment—is a radical form of attention, a message of our present and past co-existence, a reminder of how we are bound and affected by one another and that our relations and interactions shape who and what we are.

While grief and mourning has been revealed as an essential aspect of any politics of recognition or visibility, generating affect and the possibility for action, it must be joined with a further recognition in the form of a vigilance and an attention to the vulnerability of the living. Like the work of Butler, the philosophy of Simone Weil provides us with exemplary methods of exposure. For Weil, 'attention is the rarest and purest form of generosity' (Weil 2002, 104). Attention is a way of attuning ourselves to the suffering of the Other, making ourselves open and receptive to their experience of affliction. As Weil writes, 'attention consists of suspending our thoughts [...] our thought should be empty, waiting, not seeking anything, but ready to receive in its naked truth the object that is to penetrate it' (Weil 2002, 104). My purpose in turning to Weil is to demonstrate how her work offers ways of suspending human mastery and revealing an attentive quality to nonhuman vulnerability. Like Julietta Singh, Weil's thought involves an act of dispossession revealing vulnerable modes of perceiving. Singh's project is to work towards mastery's undoing. Weil, likewise, presents ruptures in sovereignty and outlines the possibility of living in more just and equal ways.

Weil's scholarly work and life were motivated by her search for truth and her attempts to transcend the arbitrary demarcations that segregate and divide us. Her concept of attention is a way of reading others, which moves beyond previous preconceptions and constructions of difference, revealing the subject in all its truth and vulnerability. Such a call for a radical sense of openness is fundamental to the theory of planetary vulnerability I am proposing. By adopting Weil's notion of attention within the context of energy culture, we can address not only how fossil fuels have impacted human lives but can also open ourselves to the ways it has shaped and effected the lives of the nonhuman. Yoon Sook Cha identifies 'an ethics of self-dispossession' (Sook Cha 2017, 11) in Weil's work, as she suggests that in processes such as decreation and attention there is a 'relationality that continually brings one outside oneself' (Cha 2017, 6). Here, I turn to Weil to consider how a plea to hear and recognise the suffering of the Other can be an earthly call of other creatures and the environment.

For Weil, attention also names a love for one's neighbour, a message of compassion which is arguably an inherently ecological one. As Timothy Morton has argued, 'Ecology is about radical co-existence', in that thinking ecologically 'involves becoming open, radically open – open forever, without the possibility of closing again' (Morton 2010, 18). Morton's definition of ecological thinking reveals how Weil's thought can be connected to an ecological consciousness which perceives the self as an interdependent being, subject to circumstance, and open to what exists outside itself. In her letter to Father Perrin her desire for a radical sense of co-existence is self-evident: 'I have the essential need, and I think I can say vocation, to move among men of every class and complexion, mixing with them and sharing their life [...] merging into the crowd and disappearing among them' (Weil 2013, 1). Weil's vocation is shaped by her desire to understand and to empathise with the Other. Attention leads to a kind of interpersonal merger, one which results in a divine sense of compassion. As Weil herself notes, 'a divine inspiration operates infallibly, irresistibly, if we do not turn away our attention, if we do not refuse it' (Weil 2002, 233). For Weil, pure attention consists of this divine and supernatural quality, allowing us to turn to suffering in all its unbearable forms and expose our own sense of fragility. Despite Weil situating her ethics within this predominantly human and theological sphere, her philosophical thought invites us to look beyond the human and attune ourselves to what has been excluded.

One of the ways we achieve this state of pure attention is through the experience of affliction—the endurance of a suffering so psychologically and physically severe that one no longer views themselves as a subject and instead becomes depersonalised, 'an anonymous focal point of pain' (Pick 2011, 31). It is a process where the 'I' is lost, and we become 'deprived of personality and made into things' (Weil 2002, 73). This experience according to Weil is not an altogether negative one, for to accept the blind incomprehensibility of suffering can lead to a sense of compassion. For Weil, the recognition of the finitude and vulnerability of human existence through affliction results in a closer relation to God and a deep sense of empathy for our neighbours. Although religiosity is fundamental to Weil's thinking, I am more interested in the ways her philosophical method enables us to read and understand the suffering of other humans and different species. Weil's thoughts on affliction offer a different approach to dehumanisation, one in which it can be transformed, like mourning, to develop an other-centred ethics. Weil relates her own experience of affliction to her experience as an employee in a Renault car factory. She declares that 'here you are nothing. You simply do not count. You are here to obey everything, to keep your mouth shut' (Weil 2002, 60). Weil's analysis of the roots of oppression in manufacturing work reveals how one becomes an object, deprived of agency, losing the ability to identify as a subject.

The radical sense of openness that Weil achieves through the notions of attention and affliction offers an alternative way of thinking about energy culture and its effects, one in which we recognise the earth and its creatures not as instrumental to human needs. As Pick has noted, 'attention is anti-philosophical; it does not produce arguments or truth claims about its object' (Pick 2011, 5) but is instead a way of seeing without identification. It is therefore a way of widening our perspectives and producing an ethics which goes beyond the principle of familiarity and anthropocentric frameworks. Like Butler's interpretation of mourning, affliction and attention can hold a transformative power that enables us to recognise the oppression of the nonhuman animal, the vulnerability of plant life and our ecologies.

Yoon Sook Cha suggests that a decreative ethics emerges from Weil's philosophy, which leads to an ethical bind towards the Other. Decreation for Weil can sometimes result in discovering the ultimate absence of God through the experience of affliction. As Weil writes,

> What is terrible is that if, in this darkness where there is nothing to love, the soul ceases to love, God's absence becomes final. The soul has to go on loving in the void, or at least to go on wanting to love, though it may be only with an infinitesimal part of itself. (Weil 2021, 172)

Within the endurance of affliction, however, a sense of the divine operates as a form of love from God, which manifests through an attunement to one another, an address to and care for the afflicted. As Lisa Radakovich Holsberg suggests, 'decreation allowed human beings to participate in the love of God through the loving of another' (Holsberg 2017, 63). For Sook Cha, the modalities in Weil's philosophy from attention, affliction and decreation offer a process of withdrawal, a way of removing operations of power. This produces a decreative ethics which has the following aim:

> For the decreative aim of obligation is specifically to preserve the other from harm, so as to preserve the core of his impersonal being. Accordingly, one does not exactly "do" anything, as it were. And the help that one proffers models a withdrawal of claims to mastery and sovereignty. (Cha 2017, 4)

The withdrawal of sovereignty and mastery is a key aspect of my approach to energy regimes and relations. If we are to conceive of just energy futures, how might the withdrawal of mastery become significant to move beyond the techno-utopian dreams of capitalism's management of nature? To withdraw is to become attentive to injustice, as Weil writes: 'the spirit of justice and truth is nothing else but a certain kind of attention, which is pure love' (Weil 2021, 92). Withdrawal, or what Pick later comes to refer to as 'letting be', offers a distinct form of love which is impersonal and distant. Although Weil's ambitions are fundamentally tied to the search for God, the implication of her philosophy can be towards the earth, to hear the cry of afflicted, to help heal the anonymous wounds of another.

Anat Pick's work in *Creaturely Poetics* is inspired by Weilian thought, and she looks to the transformative and positive effects of reclaiming dehumanisation. As she declares in her introductory chapter, 'I am interested in whether and how dehumanisation can be reclaimed as, at least partly, positive' (Pick 2011, 4). She introduces her concept of the creature: 'a living body – material, temporal, and vulnerable' (Pick 2011, 5). Reading through a creaturely prism is a way of rejecting the human/

animal binary, discovering a shared vulnerability across difference. For Pick, the human itself is not a neutral term, but on the contrary is shaped by structures of power where everything outside of 'man' is either reduced to a disposable body or branded as a resource for human utility or consumption. Pick introduces the notion of the creature as an opportunity to reconceive our own sense of animality, the fundamental finitude and vulnerability which we share with nonhuman and vegetal life.

Pick's creaturely poetics is grounded in corporeality and vulnerability, revealing a common embodiment with other species, which demonstrates a shared susceptibility to injury and pain. For Pick, the recognition of shared vulnerability can generate new forms of a multispecies community and compassion (Pick 2011, 69). Despite Pick's emphasis on the corporeal, however, the environment does not have to be excluded from the ethical radar of the creaturely. On the contrary, the creaturely theoretical framework allows an open approach to life which can include insects, plants and the environment itself. This open framework is not a means of subsuming all forms of life into one category but a way of recognising their nuances, differences and similarities—developing a non-hierarchised space where it is not an ethical case of *either or* but an inclusive and definitive *and*.

In order to become creaturely subjects, we must recognise our shared sense of vulnerability with all living things. For Pick, an orientation towards vulnerability offers a mode of exposure. She states that 'universal and shared vulnerability blurs species distinctions since humans and nonhumans alike are subject to natural law, to injury and death' (Pick 2018, 2). Recognising our finite and temporal condition can enable us to form an alliance with plant life and nonhumans as the perception of our shared state of precarity leads to an illumination of our co-dependency and our responsibility to allow other creatures to not only survive but flourish. Pick's focus on vulnerability becomes a way of breaking down the violent borders of humanity, removing arbitrary constructions by revealing similarity and adding nuance to our differences; she is guided by Weil's thought in the sense that she 'would like to make it, not less painful, only clearer' (Weil 2013, 1). Like Weil and Butler, Pick is concerned with how the exposure to vulnerability can lead to ethical action. Both Weil and Pick are concerned with non-vocal expressions of pain and suffering, that which is often unexpressed or too severe to articulate, or with no common language to express it.

The often inexpressible nature of pain is articulated by Elaine Scarry:

> When one hears about another person's physical pain, the events happening within the interior of that person's body may seem to have the remote character of some deep subterranean fact, belonging to an invisible geography that, however portentous, has no reality because it has not yet manifested itself on the visible surface of the earth. Or alternatively, it may seem as distant as interstellar events referred to by scientists who to us mysteriously of not yet detectable intergalactic screams of very distant Seyfert galaxies, a class of objects within which violent events of unknown nature occur from time to time. (Scarry 1987, 4)

Scarry highlights the difficulty in understanding and empathising with another human being's suffering; the nonhuman is not her focus, but her argument suggests the difficulty of recognising pain beyond oneself where there is no familiar schema to rely on or an immediate acknowledgement of a shared condition. Pick perceives such inability to recognise bodily pain in other creatures as a process of deflection, a term she borrows from Cora Diamond. Deflection, Diamond contends is 'what happens when we are moved from the appreciation, or attempt at appreciation, of a difficultly of reality to a philosophical or moral problem apparently in vicinity' (Diamond 2003, 57). It is a process of refusal, an inability to recognise and perceive the other as someone or something as familiar. Pick's return to the corporeal reality then is perhaps a method of communion and connection; it demands the communication of bodily vulnerability and the recognition that 'like all living beings: animals are temporal and finite: they are born, they live, and they die' (Pick 2018, 2). It is this 'rawness of nerves', a bodily affective response, which allows us to register extractive violence and its impacts on nonhuman life and the environment.

Scarry's understanding of the impossibility of communicating inner pain and suffering allows us to return to Weil when she emphasises the need to hear the silent cry. The creaturely, then, informed my Weil's theology and philosophy, is at the limits and precipice of language. As Sook Cha describes, it is a non-cognitive modality that is at work when we recognise the often silent utterance and sounds of suffering. It is here I suggest an earthly vulnerability is possible, one that moves beyond linguistic pleas to the corporeal recognition of vulnerability.

Butler, Weil and Pick all focus on the negative consequences of vulnerability, such as mourning, pain and death, and how these may lead to ethical recognition and action. As Pick suggests,

> Vulnerability dispassionately denotes the condition of being embodied as necessarily limited, and limited by necessity, but always already encompassing the dialogic relation between bodies that underlies caring. Within this vulnerable range are possible all kinds of experiences. (Pick 2011, 15)

Pick notes how there are a variety of possibilities that come with focusing on our corporeal vulnerability, experiences which can lead to ecologies of care. Often, however, vulnerability does not lead to ethical action but to a state of anxiety and a reassertion of humanity in the face of suffering. Pick explores this within the context of the Holocaust, declaring that 'cultural anxiety over species identity determines the ways the Holocaust is and is not represented' (Pick 2011, 24). The Nazi genocide of the Jewish people is shaped by a rhetoric of dehumanisation, denying their identity, and reducing them to the status of objects. In the face of such brutality, Pick highlights how literature in the aftermath of the traumatic events lead to scholarship which focused on the reclamation of humanity in works such as Primo Levi's *If This is a Man*. As Pick goes on to state, the goal is to reassert human dignity in the face of atrocity (Pick 2011, 24). Pick explores the imperatives of reclaiming humanity after an exposure to unbearable vulnerability. Nevertheless, the unravelling of the human identity, in its 'doing and undoing', reveals it as 'a tenuous category, a fragile construct' (Pick 2011, 27). Unlike Butler, Pick is interested in reclaiming animality after the endurance of dehumanisation, understanding how atrocity can leave us stripped of the constructs we use to divide ourselves from the nonhuman animal. Inspired by Weil's thoughts on affliction, Pick explores how severe suffering can lead to philosophical and intellectual insight. The afflicted are, in the words of Weil, like 'a being struggling on the ground like a half crushed worm' (Weil 2009, 69). Pick notes how Weil conceives of such affliction and suffering as a 'divine technique, a sort of theological gateway' (Pick 2011, 31). It becomes a way of accepting the blind and incomprehensible suffering all species face, and the damage that is often felt but not yet perceived or recognised. In so doing, we can establish a sense of relational subjectivity, defined not by what we are but by the relations and the environment which surrounds us. Such a perspective has important ramifications for energy and fossil fuel culture and allows us to understand the violence it presently inflicts.

Pick's creaturely fellowship manifests as transformative action as it becomes a 'way of illuminating the relations we currently have – and the ones I believe we ought to have – with the world around us, human, animal and other' (Pick 2011, 156). Utilising Weil's notion of attention, Pick illuminates the connections and bonds that can be formed and are perhaps already in place beyond the human. For Pick, 'Fellowship is ridiculous, ungainly, carnivalesque even – but solid and unquestioning. It is rooted in bodies exposed to time and (literally the roller coaster ride) at the mercy of gravity' (Pick 2011, 188). By taking notice of our embodied reality, we find that we are not atomised creatures but are embedded within and interdependent on other species and environments. All living things are 'both appreciators and victims of a material existence' (Dorsky 2014, 17). This statement, for Pick, is where a creaturely poetics begins: a simple truth which conveys our vulnerability and our suffering as an inevitable consequence of our embodiment and existence. Recognition of one's own finite conditions, as well as those of other species, is, as Pick articulates, 'the communication of the extraordinary through the ordinary' (Pick 2011, 191).

Vulnerability, and its various manifestations in the work of Butler, Weil and Pick, offers an ability to attend to and hear the cry which can remain unarticulated and non-vocal. However, there is another concept that remains crucial for my literary analysis of energy extraction: Pick's notion of 'letting be', further inspired by Weil's philosophical thought. Pick's understanding of 'letting be' is that it is 'a conservationist impulse that honours the existence of beings and things by looking-without-devouring' (Pick 2017, 49). Pick describes how Weil repeatedly returns to the act of looking. For Weil, to gaze upon beauty is to remain distant, yet empty, a non-devouring gaze which is attentive and open. This act of withdrawal for Pick is a process of letting be, something which is developed through the act of attention. Pick notes, 'as a disposition and a technique, attention entails a relaxing of personal will and the cultivation of detachment, which allow the object to emerge more clearly' (Pick 2017, 51). This sense of detachment offers a mode of mutual exploration rather than domination or mastery. As I will later explore in the work of Julietta Singh and decolonial practice, vulnerability presents new relations which are non-masterful as we withdraw from the lure of sovereignty.

As I have illustrated, Rob Nixon and Carol J. Adams point us towards the amnesia which occurs in the face of ecological damage and the slaughter of nonhuman animals and to the disparity which exists between production, consumption and their consequences. This disjuncture is predicated on violence, the process of choosing to turn away and a demonstration of the uneven development of the capitalist world system. Weil articulates this uneven process in the following statement: 'A power, whatever it may be, must always tend towards strengthening itself at home by means of successes gained abroad' (Weil 2002, 60). The process of succeeding abroad can be aligned with Western powers' means of accumulating resources from other locations across the globe and leaving environmental devastation elsewhere and physically out of sight. There is a politics to what 'remains outside our flickering attention spans – and outside the purview of a spectacle driven corporate media' (Nixon 2013, 6). Turning to Butler, Weil and Pick, however, as a combined praxis of conveying an earthly vulnerability, becomes a way of opening our attention to what has been excluded from the spectacle-driven media and the anthropocentric narrative. Nixon's identification of the 'representational challenges' (Nixon 2013, 10) confronted by cultural theorists in the face of ecological devastation is addressed through these three exemplary thinkers of exposure. Through mourning, affect, affliction as well as the creatural thought, we find that the ecological devastation and animal suffering, at the centre of energy extraction, can become exposed and felt.

However, in order to address not only the individual affects faced in response to nonhuman suffering and environmental destruction, but it is also important to turn to the colonial legacies which shape the space of the extractive zone. Commodity production, and crude oil in particular, is immersed within the histories of colonial conquest. Drawing out the relation between extractive violence, imperial legacies and the force of humanism presents a challenge to the current capitalist system. Affect and vulnerability cannot provide structural change alone, yet combined with a decolonial analysis, we can begin to rebuild and generate hope from environmental ruin.

The Human and Its Colonial Origins

In an analysis of extractive histories and the role of nonhuman, it is thus essential to interrogate the relation of capital imperialism and the construction of the human. For Kathryn Yusoff, the projects of commodity extraction and their colonial histories are intimately bound with a certain construction of the 'human'. As Yusoff suggests, 'modern liberalism is forged through colonial violence' (Yusoff 2018, 2). Such a statement is indicative of the ways humanism as a 'civilising' project is imbricated within colonial ideologies and violent histories of displacement and exploitation. Yusoff further outlines the category of the human and its counter category, the inhuman, describing how they are 'historically relational to discourse of settler-colonial rights and the material practices of extraction' (Yusoff 2018, 2). Cara New Daggett also importantly notes how this construction of the 'human' did not begin with fossil fuel industrialisation and extraction but much earlier as theorised by postcolonial theorists such as Sylvia Wynter, Walter Mignolo and Irene Silverblatt (Daggett 2019, 133). For theorists such as Wynter, this idea of Western man defined against a racialised Other originates with the Renaissance and the Spanish colonisation of the Americas. Wynter suggests that the description of the rational, political subject, the white Western Man forges the parallel invention 'Man's Human Others' (Wynter 2003, 313). Working with Wynter's ideas of the Coloniality of Being, Alexander G. Weheliye argues that for Wynter 'the promise of black studies – the numerous other ruptures precipitated by the 1960s – lies in its liminality, which contains potential exit strategies from the world of man' (Weheliye 2014, 28). Weheliye's project in *Habeas Viscus* is to attend the centrality of race to notions and constructions of the human. Like Wynter, he is interested not in conforming to the master code of Man, the colonial construction of the liberal subject, but rather to forge new and different genres of the human.

Opening Julietta Singh's *Unthinking Mastery* is an epigraph from Weheliye's work: 'What different modalities of the human come to light if we do not take the liberal humanist figure of Man as the master-subject but focus on how humanity has been imagined and lived by those subjects excluded from this domain' (Weheliye 2014, 8). In the same fashion, I aim in this book to reject the liberal humanist figure of Man who is bound to an extractive and colonial paradigm; by rejecting this construction of the 'human', we can encourage what Weheliye suggests as new modalities of the human. In Pick's terms, it can generate a creaturely poetic, rediscovering humankind outside of its oppressive delineations. Rediscovering a sense of animality, or as Pick suggests, even the vegetative aspect of ourselves, can encourage new formations of the human outside of the colonial extractive paradigm. If, as Yusoff argues, 'the "Age of Man" is a dominant and dominating mode of subjectification – of nature, the non-Western world, ecologies and the planet' (Yusoff 2018, 54), it is important to configure what Weheliye describes as exit strategies from this domain.

Like Butler, Weil and Pick, Singh describes possible exit strategies through the modality of vulnerability. For Singh, such conditions of vulnerability can begin through the process of failing to master. As Singh describes,

> In failing to master, in confronting our own desires for mastery where we least expect or recognise these desires, we become vulnerable to other possibilities for living, for being together in common, for feeling injustice and refusing it without the need to engage it through forms of conquest. (Singh 2018, 21)

By turning to narratives of mastery (which all narratives of extraction are), we discover how these narratives 'are always fragile, threatened, and impossible' (Singh 2018, 18). Vulnerability as simultaneously a passage to agency and a non-masterful form of politics can offer a new way of forging energy relations outside of the extractive paradigm. For Singh, vulnerable modes of listening, of attentiveness, can produce what she terms 'dehumanism promise', which is to become 'sensitive to those human and inhuman beings that we currently conceive as proximate to us, and most urgently to those who which still imagine as radically distinct' (Singh 2018, 64). In this sense, dehumanism can be likened to the project of the creaturely, discovering the relations we may have lost and the connections in which we ought to have. Vulnerability, however, can lead to the reassertion of yet more violence, but it also generates the possibility of care, of rediscovering relations, which foster a multispecies ethics that has significant ramifications for our energy present and future.

Vulnerability in the Extractive Zone

Singh suggests, 'reading literature can be a crucial vantage point from which to rethink the human as a product shaped and enforced through narratives that are historically, socially, politically, and filially produced' (Singh 2018, 108). The project of this book is defined by discovering different modalities of the human through vulnerability, and the possibilities and promise of such a project for our relations to energy and the wider environment. In moments of violent extraction, from the harpooning of the whale for the purpose of extracting energy, the offshore oil extraction site, to the rendering of flesh to meat, to the violent consequences of automobility, I look to the affective powers of an earthly vulnerability as a site of resistance.

In Pick's analysis of Elizabeth Bishop's poem, 'The Fish', she suggests that vulnerability arises 'as both a state of conflict and designates both violence and resistance' (Pick 2018, 418). For Pick, the fish's desperation for survival and visible vulnerability marks an act of resistance. As Pick highlights from this poem, 'fish vulnerability and resistance go hand in hand' (Pick 2018, 419). As Pick goes on to conclude, 'vulnerability is not the absence of power, but the product of power relations' (Pick 2018, 422). Magnifying vulnerability in the extractive zone presents the ubiquity of power relations yet also their possible suspension. What new ways of conceiving energy, how we interact with it and how we produce it, can become possible if there is a suspension of violent 'human' dominion and mastery? It is in the space of violent extraction in which I look for the surprising and often suppressed forms of resistance, ignited by revelations and exposure to vulnerability. By mapping the extractive conditions of capitalism, and illuminating the force and history of colonial legacies, while turning to theorists who invoke an earthly vulnerability, I hope to gesture and point towards the implications of a non-anthropocentric ethics in field of energy culture, one that holds the possibility for just environmental change.

References

Adams, Carol J. 2020. *The Sexual Politics of Meat: A Feminist-Vegetarian Critical Theory*. Twentieth anniversary edition. Bloomsbury Revelations Series. London: Bloomsbury Academic, An imprint of Bloomsbury Publishing Inc. https://doi.org/10.5040/9781501312861.

Butler, Judith. 2006. *Precarious Life: The Powers of Mourning and Violence*. Verso.

Butler, Judith. 2016. *Frames of War: When Is Life Grievable? (Radical Thinkers)*. Verso.

Butler, Judith. 2020. *The Force of Nonviolence: An Ethico-Political Bind*. London New York: Verso.

Cha, Yoon Sook. 2017. *Decreation and the Ethical Bind: Simone Weil and the Claim of the Other*. New York: Fordham University Press.

Cunsolo, Ashlee. 2017. Climate Change as the Work of Mourning. In *Mourning Nature: Hope at the Heart of Ecological Loss and Grief.* McGill-Queens University Press.

Daggett, Cara New. 2019. *The Birth of Energy: Fossil Fuels, Thermodynamics and the Politics of Work*. Durham, NC: Duke University Press.

Diamond, Cora. 2003. The Difficulty of Reality and the Difficulty of Philosophy. *Partial Answers: Journal of Literature and the History of Ideas* 1 (2): 1–26. https://doi.org/10.1353/pan.0.0090.

Dorsky, Nathaniel. 2014. *Devotional Cinema*. Tuumba Press.

Gould, Deborah Bejosa. 2009. *Moving Politics: Emotion and Act up's Fight Against AIDS*. Chicago: University of Chicago Press.

Haynes, Melissa. 2017. Animals and Energy. In *Fueling Culture: 101 Words for Energy and Environment*. Fordham Press.

Holsberg, Lisa Radakovich. 2017. Decreation and the Creative Act: Simone Weil and Nikolai Berdyaev. In *Simone Weil and Continental Philosophy*. Rowman and Littlefield.

Morton, Timothy. 2010. *The Ecological Thought*. Cambridge, MA; London: Harvard University Press.

Nixon, Rob. 2013. *Slow Violence and the Environmentalism of the Poor*. First Harvard University Press paperback edition. Cambridge, MA; London: Harvard University Press.

Out of the Woods Collective 2020. *Hope against Hope: Writings on Ecological Crisis*. Brooklyn, NY: Common Notions.

Pick, Anat. 2011. *Creaturely Poetics: Animality and Vulnerability in Literature and Film*. New York: Columbia University Press.

Pick, Anat. 2017. Nothing Now but Kestrel': Simone Weil, Iris Murdoch and the Cinema of Letting Be. In *Iris Murdoch Review*. Kingston University Press.

Pick, Anat. 2018. Vulnerability. In *Critical Terms for Animal Studies*. Chicago University Press.

Plumwood, Val. 1995. *Human Vulnerability and The Experience of Being Prey*. Quadrant Magazine Ltd.

Roy, Arundhati, David Barsamian, and Naomi Klein. 2004. *The Chequebook and the Cruise Missile: Conversations with Arundhati Roy*. London: Harper Perennial.

Scarry, Elaine. 1987. *The Body in Pain: The Making and Unmaking of the World*. New York: Oxford University Press.

Sedgwick, Eve Kosofsky. 1993. *Tendencies*. Duke University Press. https://doi.org/10.2307/j.ctv11hpmxs.

Singh, Julietta. 2018. *Unthinking Mastery: Dehumanism and Decolonial Entanglements*. Durham; London: Duke University Press.

Szeman, Imre, and Jennifer Wenzel. 2021. What Do We Talk about When We Talk About Extractivism? *Textual Practice* 35 (3): 505–523. https://doi.org/10.1080/0950236X.2021.1889829.

Weheliye, Alexander G. 2014. *Habeas Viscus: Racializing Assemblages, Biopolitics, and Black Feminist Theories of the Human*. Durham: Duke University Press.

Weil, Simone. 2002. *Gravity and Grace*. Routledge.

Weil, Simone. 2009. *Waiting For God*. Harper Perennial Modern Classics.

Weil, Simone. 2013. *A Letter to A Priest*. Routledge.

Weil, Simone. 2021. *Simone Weil: An Anthology*. Penguin Classics.

Wynter, Sylvia. 2003. Unsettling the Coloniality of Being/Power/Truth/ Freedom: Towards the Human, After Man, Its Overrepresentation--An Argument. *CR: The New Centennial Review* 3 (3): 257–337. https://doi.org/10.1353/ncr.2004.0015.

Yusoff, Kathryn. 2018. *A Billion Black Anthropocenes or None*. Forerunners Ideas First from the University of Minnesota Press. Minneapolis, MN: University of Minnesota Press.

CHAPTER 3

Oil and Meat Production Magnifying the Peripheral

Modern energy production is informed by a politics of visibility and produces what Rob Nixon refers to as representational challenges, in which deliberate corporate evasion, spatial distancing and cognitive dissonance shape both meat and oil culture. Meat and crude oil are both energy commodities, one for fuel for the body and sustenance, where the other is used for transportation and mobility, as well as other commodity derivatives such as plastics, electricity generation, asphalt and more. Their production occurs in offshore and rural peripheries, yet both are seemingly ubiquitous resources, everywhere in multiple forms and derivatives. To render visible these commodity processes, I bring together meat and oil culture by exploring two novels: Michel Faber's *Under the Skin* and Nnedi Okorafor's *Lagoon*. I argue that these texts provide a 'telescoping function' (WReC 2015, 17) by magnifying the destruction and suffering which occurs at the periphery. Turning to different types of 'extractive zones', the land itself and the body of the animal, Faber and Okorafor's narratives explore terrains of violent production while introducing voices and moments of resistance. Following Macarena Gómez-Barris, who questions 'what cultural and intellectual production makes us see, hear, and intimate the land differently' (Gómez-Barris 2017, xx), I turn to these narratives which expose the conditions of extractive capital yet offer a challenge to its present hegemony.

J. Taylor, *The Nonhuman Narratives of Energy*, Palgrave Studies in Animals and Literature,
https://doi.org/10.1007/978-3-031-94332-4_3

Before my literary analysis, I foreground the relationship between liberal humanism, colonialism and capitalism while exploring the systematic erasures they produce. I suggest how the boundary of the human is an artificial product of constant negotiation, one often informed by colonial rhetoric and practices. Following Kay Anderson, I iterate how 'humanity is not an essence, but a shifting mode of being' (Anderson 2012, 2) and that this shift between the human and the inhuman is intrinsically bound to the project of colonial extraction. The affinity between humanism and colonialism is aligned with the wider economic project of capitalist imperialism, addressing how empire and capital are related through projects of extraction. By liberal humanism, I refer to an understanding of the human subject as espoused by the Enlightenment project and political philosophers such as John Locke, who, as Julietta Singh suggests, foregrounds the modern subject within the capitalist economy, through mastery and property ownership (Singh 2018, 13). The sovereignty of the 'human', the reign of colonial extraction and capitalist ideology results in what Yusoff terms 'deadly erasures' (Yusoff 2018, 9), a subjugation 'of nature, the non-Western world, ecologies, and the planet' (Yusoff 2018, 54). Turning to these systematic and structural erasures, I demonstrate the ways Rob Nixon's concept of slow violence and Carol J. Adams' notion of the absent referent are theoretical tools which point towards the absence of nonhuman suffering and environmental destruction.

To address and explore these forgotten histories and peripheral spaces, I open my literary analysis with Nnedi Okorafor's *Lagoon*, which is set within the petroleum landscape of Lagos, Nigeria. My intention is to magnify the peripheral scenes of offshore oil extraction, the interaction of petroculture and the nonhuman, and the colonial legacies which manifest within the text through alien invasion. The novel itself traces the adventures of the biologist Adaora, and two other characters, amongst the chaos of an alien invasion in Lagos. As the city is upturned, Okorafor introduces a non-anthropocentric world of spider narrators, swordfish attacking oil pipelines, and shapeshifting aliens. This world illuminates the collision of the nonhuman and petroculture, exposing the lives impacted and affected by regimes of extraction.

Moving onto Michel Faber's *Under the Skin* I address the space of the factory farm, drawing the text into dialogue with the petroculture of Okorafor's *Lagoon*. In Faber's novel, the colonial interloper is the alien industry of Vess Corporation and their labour representative Isserley. Faber's strange and interesting reversal, however, is that this alien species, who appears characteristically in animal form, with arched backs and fur, define themselves as 'human', and what we recognise to be human life are redefined as vodsels, a meat delicacy this alien life consumes. The novel follows the character of Isserley and traces her complex relation to her role in the meat industry.

In the final stages of this chapter, I illuminate an ethics and politics of vulnerability in the texts, addressing the ways in which an exposure to vulnerable life might provide an alternative to the colonial extractivism of energy and meat production. I draw on Anat Pick's conception of the creaturely and Judith Butler's definition of vulnerability to explore these scenes of vulnerable exposure. Butler and Pick both point to a recognition of corporeal vulnerability that is at once 'perilous and enabling' (Butler 2012, 1). The encounters with an alien and animal alterity in the texts provide invitations for both violence and care. My central question is whether an exposure to vulnerability can provide a method and a mode of resistance to energy imperialism, a way to create ruptures in the capitalist system of production.

Capital Imperialism Erasure and Liberal Humanism

We have already discussed the politics of visibility that is inherent to a combined analysis of animal studies and oil culture, and the ways operations of displacement and corporate evasion shape their production processes. To expand this further, we need an analysis of how capitalist imperialism is modelled on exclusionary understanding of the human. By unpacking the ways liberal humanism is entangled in the movements of capital imperialist processes, it will thus inform my readings of Nnedi Okorafor's *Lagoon* and Michel Faber's *Under the Skin*. Explicitly, I look to how capital's deadly erasure of life, its process of peripherialisation, is tied to a particular Enlightenment understanding of the 'human'. Understanding how humanity can be a shifting mode of being allows us to explore how it can be utilised as a way to recognise lives and disregard others in the name of past, present and future extractive projects. The human subject who possesses reason and agency has often been a hegemonic tool that serves to forge a distinction between 'civilisation' and the 'rest', forming a realm of subjugation and erasure. Humanism has thus been an instrumental tool for colonial ideologies and capitalist expansion.

In Kathryn Yusoff's crucial intervention into the debate surrounding the origins of the Anthropocene, she suggests that 'modern liberalism is forged through colonial violence' (Yusoff 2018, 2). Such a statement indicates that liberal humanism, with its ideals of freedom and progress, has a counterpart of dehumanisation and violence. Yusoff's argument in *A Billion Black Anthropocenes or None* traces the history of colonialism and slavery within the geological formation and coining of the Anthropocene. Without a recognition of geology's racial inscription, she suggests the coining of the Anthropocene is a dangerous and 'deadly erasure, a rebirth without responsibility' (Yusoff 2018, 8). Yusoff's analysis of how life becomes extractable matter under colonial regimes of extraction and production provides a useful insight into the complex relationship between humanist discourse, colonialism and the capitalist project. As Yusoff goes onto note,

> Geology is a category and praxis of dispossession. It has determined the geographies and genealogies of colonial extraction in a double sense: first in terms of settler colonialism and the thirst for land and minerals, and second as a category of the inhuman that transformed persons into things. (Yusoff 2018, 68)

The categorical distinction between human and inhuman is an important axis that determines life—humans, nonhuman animals and land—as extractable matter. Yusoff suggests that it 'facilitated the division between subjects as humans and subjects priced as flesh' (Yusoff 2018, 9). Here, Yusoff's focus is on slavery and the reliance on black lives as the source of the labour necessary for colonial projects of extraction. I would argue that her questions around the role of material agency, 'outside the cozy structures of humanism' (Yusoff 2018, 9), also have an acute relevance for a non-anthropocentric politics and can be used to address the inadequacy and violence which emerges from the human and nonhuman binary.

The human and the nonhuman divide is highlighted by Yusoff as a key component underpinning colonial regimes, and she notes that 'both enslaved land and ecologies become subject to encoding as inhuman property as a tactic of Empire and European world building' (Yusoff 2018, 68). Colonialism is thus a project of material and discursive dehumanisation, one which turns enslaved people, colonised subjects, and land into property and matter, excluded from the privileged category of the human. This negotiation to establish the borders between what is human and what is not reveals how the human itself becomes a fragile concept, appointed according to existing power structures and complicit in colonial ideologies. Dehumanisation occurs then in the name of humanism, a reigning power structure which denies certain lives their existence, rendered extractable tools and property. As Yusoff suggests, just 'as land is made into tabula rasa, so too do Indigenes and Africans become rendered as a writ or ledger of flesh scribed in colonial grammars' (Yusoff 2018, 33). The colonial project reshapes lives, bodies and ecologies, into these grammars transmuting life into inert resources for capitalist trade and commerce.

For theorists such as Franz Fanon, this process of dehumanisation is central to colonialism. The 'Man of Reason' along with the many different conceptions of man: 'scientific, religious, political – have always, as Fanon constantly reminds us, entailed the dehumanisation of certain categories of individuals, with genocidal results' (Alessandrini 2009, 80). Fanon's remedy for those who have been denied subjectivity is however to proclaim a renewed humanism 'predicated upon a formal repudiation of the degraded European form' (Lazarus 1999, 189). Although Fanon provides a critique of the traditional model of the human subject and its connection to colonial regimes, he reasserts the need for a category itself to restore and reclaim agency.

Inspired by the work of Fanon and many other critical race theorists and philosophers, Sylvia Wynter also critiques the sovereign subject of humanism and traces its history and connection to colonialism. For Wynter, the local concept of Man/Human has an imperial universality. As she notes, 'the idea of Man at a particular moment of world history, the European Renaissance, was also the foundational step for building racism as we know it today' (Wynter 2003, 118). In order to think and live decolonial practice, Wynter's work suggests it is necessary to unsettle and undo 'Western conceptions of what it means to be human' (McKittrick 2015, 2). By unveiling the coloniality of human identity, the process reveals the often-unacknowledged violence of humanism, exposing its exclusions and hierarchical structures.

Humanism, however, not only provides the justification for colonial settler rights and racist violence, but it is simultaneously linked to 'material practices of extraction' (Yusoff 2018, 2). Central to humanism and its exclusions is what Yusoff calls the 'afterlives of geologies [...] the indigenous dispossession of land and sovereignty in the invasion of Americas through the ongoing petropolitics of settler colonialism' (Yusoff 2018, 3). The afterlives of geology are regarded as inhuman matter, the exposable externalities to the capitalist imperial project. Humanism and colonialism are thus connected to the regime by which capital transforms life into commodity. Colonialism and capitalism are both projects of dispossession and dehumanisation rendering colonised subjects and land into the grammars of capital. The political project of colonialism, one of exclusion, domination and hierarchy, is historically embedded in the material and economic project of capitalism.

Within the context of contemporary energy production, while no Western corporation rules directly through colonial power, the legacies of colonialism exist through these same regimes of capital and economic imperatives. For Ellen Meiksin Wood, there is a lack of transparency within capitalist imperialism, ensuring that 'it is much harder than it was in earlier colonial empires to detect the transfer of wealth from weaker to stronger nations' (Wood 2005, 4). Wood suggests that compulsion and obedience are not imposed through a ruling nation, but rather by the demands of the market. By turning to the market flows of energy within both oil and meat culture, we can magnify this capitalist imperial operation and therefore expose its violence.

Capitalist imperialism moves through a physical and metaphorical erasure, moving production to semi-peripheral and peripheral locations, prioritising the economic and thus transmuting life—human and nonhuman—into commodity. Like the clear and violent divide between the human and animal, capitalism foregrounds hierarchical dichotomies which shape subjectivity and turn persons, as well as other species into 'things'. To explore these 'deadly erasures' (Yusoff 2018, 8) is to recapture the forgotten histories of bodies and geologic matter utilised by and absorbed into an anthropocentric, as well as a white male-dominated world.

Such erasures recall Nicole Shukin's *Animal Capital*, which describes the historical amnesia of the origins of the manufacturing Ford assembly line production, noting how it was modelled on the disassembly of animal bodies in the slaughterhouses of Chicago. This amnesia is perhaps symbolic of the contemporary power relations found in energy production, which presents a façade of technological expertise, a controlled and distant form of management—systematic erasures enabling limitless consumption and forestalling of ethical considerations, turning bodies into resources, to inert and passive matter.

Returning to Adams' concept of the absent referent, it is clear how the process of turning bodies into commodities is illustrated by this term. In the opening of the 1990s edition, she dedicates the text to the anonymous animals slaughtered by the meat industry. As outlined in the second chapter, the central concern of Adams' text is absences, the ways in which the meat on our dinner plates bears little relation to the animal who once lived and remains absent from the process of consumption. Adams' acknowledgement of this severance between consumption and the production process gestures to a critique of capitalism itself, the limitless consumption of which is enabled through absences and erasures. Like Shukin (Shukin 2009), Adams describes how the Ford assembly production line mirrored the disassembly of animal bodies in Chicago slaughterhouse. Adams goes onto describe that 'one of the basic things that must happen on the disassembly line of a slaughterhouse is that the animal must be treated as an inert object, not as a living, breathing being' (Adams 2020, 53). In such cases, it could be argued that it is humanism and its hierarchical relations which permit such a division—an ethical and affective blindness to what appears before you.

However, what if we were to extend Adams' notion of the absent referent to think about how the oil industry likewise moves through systematic and structural effacement and elision, therefore bringing meat and oil culture into dialogue? A key component of Adams' absent referent is the need to address the ways in which language shapes animals into a product. How is crude oil similarly transformed by a language of commerce which detaches it from a previous history in which it was once apart? Oil is often cast as the invisible actor within military conflicts and invasions while the effects of extraction leave little trace within the cultural and social imagination. If Adams' project is to make what is absent present, how can we embark on a project of exposure to make oil and its effects visible, whilst also situating fossil fuel extraction in dialogue with meat production? To do so, the feminist-vegetarianism Adams advocates could have ramifications for energy relations as well as the mass consumption of meat products.

Significantly, Adams opens Chapter Five, 'Dismembered Texts and Dismembered Animals', with a quote from Simone Weil:

> Documents originate among the powerful ones, the conquerors. History, therefore, is nothing but a compilation of the depositions made by assassins with respect to their victims and themselves. (Weil 2021, 219)

Weil's words recalls Yusoff's project of situating the Anthropocene within the history of slavery, plantations and colonialism. As Yusoff goes on to note, 'the birth of a geologic subject in the Anthropocene made without examination of this history is a deadly erasure' (Yusoff 2018, 8). Addressing the history of extraction and current consumption likewise must address the lives upon which it depends—the nonhuman bodies, the organic matter, the creatures both living and dead which form the life-blood of energy systems both as food and fuel.

Yusoff suggests that extractable matter must be passive, awaiting activation through the mastery of white men (Yusoff 2018, 3). Like Adams, she reiterates how living beings are turned into extractable objects performing a new and different function co-existent with and controlled by economic and market imperatives. This co-existence is made possible through a clear and didactic assertion of humanism, a colonial white ruler, and the erosion of the histories and lives which have been forced into the process of production and extraction. In order to challenge what Adams' coins 'patriarchal consumption' (Adams 2020, 186), which is powered by Yusoff's notion of 'humanism as erasure' (Yusoff 2018, 57), I explore not the master narrative but the semi-peripheral and peripheral stories to be found within spaces of extraction and production.

To turn to the periphery and its relation to extraction is partly an attempt to visualise and narrate Nixon's notion of slow violence. Like Adams, Nixon is concerned with absences. He draws our attention to the fact that 'politically and emotionally, different kinds of disasters possess an unequal heft' (Nixon 2013, 3). The slow emerging catastrophes of ocean acidification, deforestation and the dumping of toxic waste, for example, rarely receive the attention of the calamities of burning buildings, war zones and 'terror' attacks. Although both Adams and Nixon are dealing with different forms of oppression and destruction, they both address the challenge of visibility, the difficulty of representing what is often obscured from view. Positioning them in dialogue, however, connects both environmental and animal ethics, which at times can be in tension with one another. In Nixon's introduction to slow violence, he cites Rachel Carson's *Silent Spring* as inspiration for his concept—the idea of 'death by indirection'. *Silent Spring* is also a useful text to understand the indirect violence and suffering of the nonhuman as well as environmental damage. In the case of both animal suffering and environmental catastrophe, there is a sense of anonymity—a structural form of violence where accountability is difficult to locate. This also points us to an absence, one enabled by the capitalist system which moves through erasures, eliding different lives and their histories while removing responsibility into anonymous focal points. The movement of slow violence functions through the influence of Western powers and their economic and colonial dominance. In relation to petro-capitalism, Nixon notes that 'Western multinationals exerted a disproportionate influence over the terms of extraction with their third world state partners, inhibiting democratic dispensations from developing while exploiting environmental, health, and labour climate far more lax than the legislative controls corporations were subject to back home' (Nixon 2013, 71). In such cases, we witness the effects of the division between imperial cores and colonised peripheries, the progress and development of some nations through its extraction from others.

Under the Skin and *Lagoon* explore extractive zones through images of the surreal, fantastic and the alien, yet in both narratives they sharply capture the uneven conditions of extractive capital. Both locations of these literary works, Scotland and Nigeria, have been subject to petro-capitalism and oil imperialism, in that other nations and multinational companies have benefitted from their mineral wealth. Turning to these peripheral and semi-peripheral locations and their unlikely literary protagonists allows for a different side of the conditions of under development and exploitation to be narrated, a story of extractive capital from the nonhuman peripheries.

Poetics of the Offshore and Colonial Legacies in *Lagoon*

Nnedi Okorafor's *Lagoon* explores the ramifications of an alien invasion in the city of Lagos, Nigeria. The narrative presents connections between the alien interlopers, the legacies of colonialism and the present regimes of extraction which dominate the Nigerian landscape. *Lagoon* thus produces an imaginary of the offshore revealing 'sights of the unseen' (Nixon 2013, 15), conjuring images of abandoned oil rigs, burst pipelines and mass roadkill. She provides the essential 'telescoping' (WReC 2015, 17) function, magnifying the peripheral space and subjects within the petroleum landscape. The focal point of my analysis of *Lagoon* will thus be underpinned by Nixon's notion of 'sights of the unseen' (Nixon 2013, 15). In his work, Nixon suggests that,

> Writer-activists can help us apprehend threats imaginatively that remain imperceptible to the senses, either because they are geographically remote, too vast or too minute in scale, or are played out across a time span that exceeds the instance of observation or even the physiological life of the human observer. (Nixon 2013, 15)

Okorafor provides this attentiveness to the remote space and the minute scale: from the offshore oil site to pipelines bursting into the habitats of marine life. *Lagoon* traces the cascading violence of extractive industries while attending to the creatures who are entrapped within its infrastructure. Okorafor provides a vision of the human and nonhuman lives who exist within the marginal spaces of production and extraction. My analysis thus offers a perspective and magnification of who and what resides within the margins and peripheries. My literary readings focus on different episodes in the narrative which reveal the collision between nonhuman life, petroculture and its infrastructure. I also further address the connections between contemporary regimes of extraction and their connection to the legacies of colonialism, illuminating the ambivalent role an alien invasion holds within the Nigerian context.

Removed from Western economic centres, Okorafor explores the consequences of an alien invasion at the periphery, an extraction and production zone, in the city of Lagos. Nigeria's history is entrenched in the legacies of colonial and capitalist violence. Rich with natural resources, Nigeria has been a location for Western exploitation of neo-colonial and capitalist enterprise. We can illuminate Nigeria's economic placement through world system theories of combined and uneven development, where Nigeria in many ways has been 'sacrificed for the development of others, such as the countryside for the city […] the colony for the metropole, or even one city for another' (Harootunian 2000, xv). Despite the country's independence in 1960, the force of NGO's and capitalist enterprises, enforced austerity of National Debt from the World Bank, and the extraction of valuable natural resources have resulted in a reassertion and dominance of Western industries and corporations. In this process of extraction and development, Marxist group, Warwick Research Collective, note that,

> Capitalist development does not smooth away but rather produces unevenness […] the face of modernity is not worn exclusively by the futuristic skyline of the Pudong District in Shanghai or the Shard and Gherkin buildings in London; just as emblematic of modernity as these are the favelas in Rocinha and Jacarezinho in Rio and the slums of Dharavi in Bombay and Makoko in Lagos […] These constitute the necessary flipside of the mirroring opacities of postmodern topos like the Portman Bonaventura Hotel. (WReC 2015, 13)

It is amidst these uneven conditions, the peripheralised space of maldevelopment of modernity that Okorafor sets the narrative. Okorafor explores the implications of an alien invasion outside the Western and economic metropolis, where the question that arises is who, beyond its borders, will notice? A location where, in many ways, the science fiction apocalypse and alien invasion has already taken place.

What Jennifer Wenzel terms the 'petro-promise' (Wenzel 2006, 212), 'the illusion of a completely changed life, life without work, life for the free', is presented to the people of Nigeria with the discovery of commercial oil deposits on the eve of Nigeria's independence (Wenzel 2006, 211). The reality, however, is of a different nature as the discovery of crude oil resulted in wealth for a minority, a process of underdevelopment, and poverty for the majority of the region. As Michael Watts states, 'most Nigerians are poorer today than they were in the late colonial period' (Watts 2014, 44). Wenzel describes this process as one in which 'development begins and turns backwards' (Wenzel 2006, 213). Okorafor's narrative points to these polar extremes of urban capitalist development alongside exploitation and poverty through the alien invasion, which offers the promise of technological innovation and wealth, yet the utopian outcome is perhaps never quite realised. The alien invasion in many ways is resonant of a history entrenched within colonialism and neo-colonial control.

Okorafor's science fiction text, however, subverts the dominance of white Western hegemony through her development of Africanfuturism. Previously, Okorafor defined her work as Afrofuturism, a growing cultural and aesthetic movement which focuses on the intersection of African diaspora, technoculture and sci-fi. Okorafor claims her literary work can be placed in what she terms as either Africanfuturism or Africanjujuism, which is distinct from Afrofuturism by being directly rooted in African culture, mythology and history. Instead, she suggests Africanfuturism does not privilege or centre the West. In moving beyond Western centrism, to foreground African history and culture, Okorafor magnifies what is constructed as peripheral space and experience. By doing so, Okorafor also challenges the colonial and neo-colonial imaginaries of Western SF (science-fiction), where outer space has often been a site for a literary landscape of colonial conquest, perpetuating a vision of imperialism into interstellar regions to expand empires and envision future colonies. Her subversion of colonial and neo-colonial tropes of SF allows for a new imaginary of what could be, outside of capitalist and oil imperialism, one which is connected and bound to Nigeria's history under colonialism and its ongoing legacies in relation to the dominance of multinational oil companies.

The alien invasion itself presents ambivalent figures in which critics such as Michael Paye have interpreted the aliens as hybrids as representative of Nigerian folklore advocating a new ontology and ethical approach to energy resources. The very act of invasion, however, can be understood as a violent act and resonant of the different forms of imperial and neo-imperial powers including oil companies such as BP, which exploited resources in Nigeria. This is not to say the aliens are clear allegorical figures of Western powers and the dominance of the oil industry but rather to suggest that the fictionalised event of an alien invasion exists within a continuum with the legacies of colonialism in Nigeria.

Lagoon is pervaded by scenes of ruin not simply in the wake and chaos of an alien invasion but through the environmental devastation caused by the oil industry. Kola, the young daughter of Adaora, addresses the alien they have named Ayodele and states: 'My mother says the waters are all dirty and dead because of the oil companies' (Okorafor 2016, 68). Journalist and writer Peter Maas likewise notes that 'the oil in Nigeria has not stopped leaking' (Maass 2010, 65). Most of Nigeria's oil was found in the Niger Delta, where significant extraction took place during the 1960s. Maas notes how the local people received little wealth from the mineral resources found on their land. This controversy over mineral rights, according to Maas, 'accelerated a process of national break down' (Maas, 65). He describes the extraction sites in Nigeria as a 'landscape of ruin' (Maass 2010, 65). The continuous flow of oil bursting from the extraction pipes goes relatively unrecorded in Nigeria, where it is a hazard seen as inevitable collateral damage, a form of slow violence which rarely reaches the television screen.

While researching for his book *Crude World*, Maas travelled through 'Nigeria, meeting oil executives, oil warlords and immiserated fishermen whose livelihood has been destroyed; it was a journey into the carbon hell on the other side of our gas pumps' (Maass 2010, 65). It is this journey in which we embark on through reading *Lagoon*, taking us to the production side of the petroleum pipeline. However, Okorafor's fictional account of Nigeria reveals not simply the devastation evoked from oil culture but also the vibrancy of Lagos, which is bursting with its unusual sea creatures, folklores and mythology.

Okorafor opens the narrative from the nonhuman peripheries of our world system beginning with the perspective of a swordfish and its collision with an oil pipeline. Okorafor was inspired by the news story of a swordfish attacking the Angola pipeline, described by journalist

Christopher Helman as an 'orchestrated retribution by Poseidon's creatures for BP's 2010 Deepwater Horizon disaster' (Helman n.d.). Okorafor recreates this episode but inside the swordfish's subjectivity. Frustrated with an anthropocentric world, the swordfish plummets towards the pipeline in order to vanquish its malignant presence within the ocean. As the pipeline bursts and oozes with 'black blood' (Okorafor 2016, 4), the swordfish radiates with joy at having succeeded at attacking the black snake which intrudes upon their habitat. The swordfish imagines the oil pipeline as an animate agent, a monstrous oppressor cutting through the ocean bed. Okorafor's use of a nonhuman narrator opens up possibilities of thinking and feeling beyond the human. There is an emphasis on the swordfish's interiority, her experience of emotion, the anger evoked from her polluted waters from the continuous oil mining within the seabed. Okorafor opens the text with this rare and underreported news story in order to shift our perspectives from the immediate and explosive violence which proliferates across media headlines. This aims to draw our attention to the minute scales and events geographically remote: the lives impacted by oil imperialism, the human and the nonhuman lives who are not centred in the media or are denied cultural focus. By entering the swordfish's subjectivity, we immediately encounter another way of feeling and seeing, introducing a non-anthropocentric framework within petrocultures. This opening allows for this alternate perspective of petro-modernity, its consequences and effects on different life forms: extractive capitalism narrated from the nonhuman.

Okorafor also takes us to the unexplored and derelict zones of the offshore oil site. In the chapter titled 'Offshore', the character Agu is taken to an offshore oil rig. Within the narrative, there had been a report that thousands of gallons of crude oil was spilling into the sea and the production facilities abandoned due to the chaos of the invasion. Out on the boat heading towards the offshore oil rig, Agu observes:

> The rig was a spidery structure made of concrete and rusty steel. Anchored firmly to the seabed by steel beams, it was a decades-old monster, a hulking unnatural contraption of production facilities, drilling rigs and crew quarters [...] it was usually a place of noise and activity. Now it was deserted and quiet. (Okorafor 2016, 95)

The material and geographical distance associated with oil production is eradicated within this passage, as Okorofar magnifies the offshore oil site, its unnatural force and weight entrenched within the seabed. She emphasises its unnatural quality, recreating it as a monstrous entity, rupturing a natural landscape. There is a personification of the rig as a 'decades-old monster' festering and polluting the space of the offshore location. If, as many petro-critics have noted, oil and its production is uniquely occluded, *Lagoon* forces us to recognise its presence. Sheena Wilson notes this absence and evasion of oil within our visionary field suggesting that 'not only does its exchange value engender an enormous corporate project of hiding, an explicit machinery of deception and spin, its pervasiveness, its presence everywhere, perhaps singularly christens its position as "hidden in plain sight"' (Wilson 2012, 4). What Okorafor exposes is that oil production is starkly visible for certain people, those on the other side of the gas pump, where it can organise and dictate their existence. For oil to feel invisible is, in some sense, a privilege, in the sense that one is detached and severed from the production process. The spectral absence of oil, to which Wilson refers, echoes the work of Adams in relation to meat production, which argues that animals are made literally absent from meat eating because they are dead. In the act of what Adams refers to as flesh eating flesh, the animal no longer exists on both an abstract and material level. In the same ways oil and its many derivatives are not associated with the environmental consequences of production, meat production is similarly shaped by a cognitive dissonance, an enforced fragmentation which dictates the roles of the consumer and the resource. Okofaror's narrative explores the effects of oil in this way, making the absent referent present.

One of the ways in which Okorafor highlights the peripheral effects of oil culture is through the predatory road called 'the bone collector' (Okorafor 2016, 119). Our first introduction to the bone collector is through the perspective of a seven limbed spider. Although the road is yet to become a conscious entity, absorbing human and nonhuman life into its gut, the spider takes a leap surging across the road in order to avoid the oncoming traffic. We witness the outcome of his venture across the road with a 'crunch' (Okorafor 2016, 120). 'The crushed body of the large seven-limbed tarantula sinks into the road's sun warmed surface' (Okorafor 2016, 120). The central character Adaora drives across the small stretch of road unaware entirely of the crushed spider's body flattened into the cement. In this moment, Okorafor illuminates the death of the supposedly insignificant, a peripheral life and a marginal incident which takes centre stage. She names the highway the bone collector because of the many remains of human bodies, broken vehicles and the animal life it has collected over its history—a strange archive of merged remains of machine, human and animal. In the later stages of the novel, the road itself becomes a monstrous entity, eating anything within its wake. The horrors and effects of the automobile culture become magnified in this surreal depiction of a mass roadkill, where the road itself becomes a carnivore consuming the flesh and bones of anything driving or passing along the highway.

In the chapter the 'Road Monster', the narrative follows a Nigerian family fleeing the chaos of the alien invasion and the rising sea waters. As they abandon their home and head for the road, they encounter the strange monstrous entity which is the bone collector. The father of the family reflects that 'these roads are full of ghosts' (Okorafor 2016, 204). The emphasis on the apparitions which haunt the Nigerian petro-infrastructure serves to envision the slow violence of oil culture—not only from the landscape polluted by oil spills which never stops leaking but from the many lives taken by a world dominated by the automobile. The road itself becomes a carnivorous machine consuming all within its reach. The terrified family hear its rumbling cry: 'OOOMM' the road said (Okorafor 2016, 207). This monstrous and alien creature is not the unfamiliar beings arrived from outer space, but rather one of human creation: the horror of petroculture is crudely awakened, the road violently consuming its own creators. In this Frankenstein-style horror story, the family smell the pervading element of blood and hot tar on the breath of the monstrous road, the voice of the road bellowing towards the family. However, an alien arrives to protect the family from this monstrous road. The road sinisterly announces, 'I have always collected bones, I am the road' (Okorafor 2016, 208). The vibration of its voice reverberates across their bodies, making them feel 'like nothing but meat, like it could shake the meat from my bones, the bones it wanted' (Okorafor 2016, 208). This experience of becoming meat strangely reverses the human and the inhuman, the feeling of now becoming edible as a challenge to human mastery.

This experience resonates with ecofeminist Val Plumwood's account of becoming prey, which I explored in the second chapter. In this account, Plumwood emphasises human vulnerability, the feeling of utter helplessness when subjected to another's appetite. She suggests that 'the illusion of invulnerability is typical of the mind of a coloniser' (Plumwood 1995, 34). Vulnerability, the experience of becoming prey, of our bodies being edible is a way of undermining this coloniser logic. Okorafor illuminates the predator inherent within colonisation and oil culture through this image of a monstrous road while also revealing the animality within the human itself as mere flesh and bone. The image further echoes the exploitative practices of the oil industry, which have now become manifest within this all-consuming road. The alien woman, however, sacrifices herself to the monstrous highway; she screams to the road, 'collect my bones and then never collect again' (Okorafor 2016, 208). The road's appetite seems satisfied with this final body, leading the father of the family to conclude that the aliens are indeed agents of change.

Like Michael Paye, critic Gemma Field has also interpreted the aliens as a force for ethical change, a movement away from colonial and neocolonial influence to alternatively embrace the origins of Nigerian Folklore to which the aliens often allude. The sacrifice the alien makes to save the people of Lagos from the horrors of the predatory road certainly lends itself to such interpretation. My reading, however, does not follow this line of interpretation. As previously mentioned, the force of invasion cannot be seen as an entirely neutral act in the context of Nigerian history. Instead of being a one-dimensional force for good, I suggest that the aliens' role is far more ambiguous. Johanna Pudnt likewise identifies the aliens' ambiguity, arguing that their arrival can be interpreted as affirming either a 'technocratic neo-colonialism or an anti-colonial awakening' (Pudnt 2018, 177). Charles O'Connell also suggests that 'despite the aliens not presenting themselves as colonisers, the enforced imposition and technology remains a neo-imperial hallmark of neoliberal developmentalism' (O'Connell 2016, 299). The aliens' imposition from the outside and their enforcement of a technocratic solution to energy undermines their position as solely utopian figures.

Oil capitalism and energy production in *Lagoon* is characterised by its colonial imagery and Okorafor's gradual unpacking of the humanist project with which it is associated. It is unclear, however, whether the alien invasion interrupts the capitalist production process or whether it instigates another form of capitalist 'innovation'—their arrival being a form of 'disaster capitalism' pushing the nation into further chaos and proposing a form of alien techno-utopianism as new approach to energy consumption. The alien Ayodele insists that her and her people 'are change' (Okorafor 2016, 39). For Michael Paye, this change is a 'short term crisis combined with a hybridisation of Nigerian mythologies, indigenous belief, with new technologies and ways of seeing' (Paye 2019, 24). The critical response to *Lagoon* overwhelmingly reflects Paye's position where the aliens are situated as a force for ethical technological innovation, a transition from fossil fuel energy and extraction.

There are references within the narrative, however, which complicate such an interpretation of these alien life forms. When the President of Nigeria meets with the rest of Ayodele's alien people, Adaora reflects that 'humanoid figures [...] reminded her of something out of star wars [...] all the creatures she saw now were whitish-blue skinned, with huge black eyes and long long arms, legs and neck. They even moved with the same fluid motions as they had in the movies' (Okorafor 2016, 251). The *Star Wars* analogy refers to the film *Attack of the Clones*, where the aliens to which she is referring are called the Kaminoans', ultra-capitalist figures who clone and produce armies to those with sufficient funds. In *Star Wars*, they are neither framed as a source of evil or a moral good. Instead, they are situated as profit-seeking entrepreneurs in the industry of war. The Kaminoans specialise in alien cloning technology; like the aliens in *Lagoon*, they have inhuman technological abilities. The aliens therefore could be illustrative of a techno-utopian outlook and advancement. Imre Szeman has highlighted such flaws in the techno-utopian perspective:

> The notion of technological evolution lies at the heart not only of techno utopian solutions to the disaster of oil but of modern imaginings of science more generally. Technology is figured as just around the corner, as always just on the verge of arriving. Innovation can be hurried along (through increased grants, for instance), but only slightly: technological solutions arrive just in time and never fail to come. (Szeman 2014, 814)

Techno-utopian discourse is demonstrated through the alien figures of *Lagoon*, a discourse which is often similarly projected by government officials that the future will look more or less the same, but with the development of innovative technology to move away from the problems of the present. The President of Nigeria in the text declares that the alien visitors 'intend to stay. And I am happy about it. They have new technology, they have fresh ideas' (Okorafor 2016, 277). Is this really a break from a capitalist mode of production? Or is it instead merely another branch of capitalist development in the form of alien technological expertise? The techno-utopianism the aliens advocate does not challenge the systematic and structural inequalities oil capitalism perpetuates. As Szeman ironically notes, 'all of our worst fears about the chaos that will ensue when oil runs out are resolved through scientific innovations that are in perfect synchrony with the operations of the capitalist economy: problem solved, without the need for radical ruptures or alterations in political and social life' (Szeman 2014, 813). The conclusion that the aliens will introduce radical new technology to save them from multinational oil companies who exploit Nigeria's mineral resources resonates with Szeman critique of techno-utopianism. Through embracing a techno-utopianism, there is no global, social and structural change within the inequalities which currently shape Nigeria. The techno-utopian worldview aims to simply relieve the issue of oil running dry and instead turning to a new source of energy, which may continue to perpetuate global and national corruption, as well as maintaining an existing extractive logic.

Okorafor's focus on both apparent peripheral spaces and peripheral lives produces a way of capturing the slow violence evoked throughout Lagos and records the impact and influence of multinational oil companies. Throughout the text, oil culture is the landscape and the backdrop to chaos of an alien invasion. By magnifying what can often be 'hidden in plain sight' (Szeman 2015, 4), Okorafor presents to the reader the story of oil extraction from the lives in which it impacts both human and animal. The encounter with an alien Other who comes from the outside cannot be entirely separated from the history of colonialism within Nigeria. Their projection of techno-utopianism is perhaps not a way of reorganising power relations but can serve to maintain the systematic and structural inequalities at the heart of petro-capitalism. Their position in the narrative is not one of complete neutrality but can be interpreted in some ways as a neo-colonial power and resonant of extractive industries. *Lagoon* offers an illumination of who exists within perilous space of the extractive zone while contextualising the colonial and neo-colonial legacies which haunt the landscape of resource-rich regions in the Global South showing the devastation at the heart of extraction regimes.

Hidden Enclosures: Peripheral Spaces and Subjects in *Under the Skin*

Under the Skin is a novel that illuminates both a semi-peripheral location, Scotland, with its history of being subjected to oil imperialism, and to the peripheral bodies, the nonhuman lives which are categorised as meat. It is through the eyes of an alien life form in which we are awakened to the hidden enclosures of animal slaughter and meat production. Faber presents the colonial interloper and extractive project through the business of Vess Industries, which is presumably a corporation from another planet. Like in *Lagoon*, the alien invasion similarly presents the legacies of colonialism, but in this case the resource to be extracted and produced is 'human' flesh. Here, using both Adams' notion of making what is absent present and Nixon's concept of the 'unseen' (Nixon 2013, 15), I illuminate the parallels between meat and oil culture and reveal the politics of visibility which shapes these regimes of production.

Like crude oil, meat is a seemingly ubiquitous resource, the animal's flesh and bone forming various different commodities. The growth of the livestock industry and the meatification of diets poses environmental problems which can parallel those created by the fossil fuel industry. The continuous expansion of peripheral production zones where animals are slaughtered also creates a severe material gulf. This gulf creates a rupture between the human and other species, where ethical transition is forestalled and inhibited. Capitalist mass meat production is shaped by such cognitive dissonance and relies on a detachment between production and consumption. This detachment is further enforced through an ideology of who is classified as 'human' and who is left to a sphere of disposability or a commodity item for profit.

Faber sets the narrative in Scotland which holds particular significance in the context of oil culture as, like Nigeria, it is a place of extraction subjected to private influence which determines who owns and receives the benefits from their resource mineral wealth. Commercially viable oil fields were found in the North Sea in 1969, where Prime Minister James Callaghan declared that 'God has given Britain her best opportunity for one hundred years in the shape of North Sea Oil' (Callaghan 2015, 2). However, as in the case with Nigeria, it was the private multinational oil companies including BP and Shell who profited from the extraction enterprise. As Christopher Harvie states, 'an energetic and unapologetic international capitalism, rooted in two less stable regions – Texas and the Middle East – had taken an ageing central stage. A Scotland whose political culture was both semi-collectivist and post-imperial found itself being "colonised"' (Harvie 1994, 6). Although Faber does not explicitly deal with this context of oil politics and colonisation, the alien corporation arriving in Scotland to extract and farm a resource holds a particular resonance within this history. In some ways, Vess Industries can be viewed as an allegory of the multinational oil companies which came to centre stage within the discovery of Scotland's oil. In Harvie's account of the North

Sea oil, he suggests there is a deliberate occlusion associated with the off-shore oil derrick. He argues that despite the oil derrick being an obvious symbol of 'industry [...] off shore oil was discreet, and geographically withdrawn from the main population centre' (Harvie 1994, 259). Rather than being a symbol of communal industrialisation and union action such as the communities arising from work in the coal mines, the oil derrick is an isolated and derelict zone.

In a similar vein, Faber opens his text within a derelict and isolated space with the central character, Isserley, cruising down the highways to search for the desired commodity. On the empty A9 road, there are scattered dead bodies, scenes of roadkill: 'furry carcasses of unidentifiable forest creatures littered the asphalt, fresh every morning, each of them a frozen moment in time when some living thing had mistaken the road for its natural habitat' (Faber 2006, 2). Like Okorafor's spectral highway, with the broken remains of human and animal bodies crushed into the cement, Faber too exposes us to unidentifiable deaths from oil culture.

The dominance of the automobile is foregrounded from the very beginning with Isserley's position in the driver's seat, searching for the ideal specimen for the meat business enterprise. She scouts the landscape sizing up the different hitchhikers she encounters. As she drives past the first hitchhiker, 'she gives herself time to size him up. She was looking for big muscles: a hunk on legs. Puny, scrawny specimens were no use to her' (Faber 2006, 1). The erotic overtones of Faber's language suggest she is on the hunt for a sexual partner; it is later revealed that her motivation is of a sinister nature, an aim to secure their bodies for a meat processing plant. The dissection and atomisation of the male human body into its parts for meat consumption echoes Adams' critique of the subjugation of women and animals. Adams stresses the abstract interchange between women and animals, where both are objectified, eroticised and viewed as items to physically or metaphorically consume. Faber's reversal, however, is that it is the male who becomes objectified and dissected into parts. As in *Lagoon*, the human is once again revealed as an edible commodity, stripped of its 'human identity' and exposed as mere flesh. This revelatory politics shapes much of Faber's narrative where he points towards habitual processes and uncovers the horrors beneath them, exposing the production process which is often forced into peripheral worlds.

The process of invisibility and systematic erasure is central to the meat industry as it is in the oil industry. Despite the absence of live animals from our everyday experience within modernity, their dead bodies proliferate in the form of different products. Like oil and its many derivatives such as plastic, the animal body is everywhere within our consumer goods. As Sherryl Vint notes, 'the use of animals in contemporary society is increasingly invisible: they are hidden away in laboratories and factory farms; slaughtered at mass disassembly plants and transformed into sanitised packages of meat; visible only in mediated forms on Animal Planet or National Geographic television, but purged from city geographies' (Vint 2010, 1). In turning to these peripheral spaces and subjects, Faber forces the reader into the space of the factory farm through the aliens' project of Vess Industries. As the novel develops, we eventually come to the underground pens where the 'vodsels' are kept for fattening. In this subterranean environment, both Isserley and the son of the owner of Vess Industries witness the horrors which are happening down below. Despite Isserley's position as a Vess Industry worker, attaining and capturing the bodies for meat processing, she has never entered the dungeon-like pens where the slaughter occurs. Like the modern consumers' own detachment from the death of animal and the meat we find upon our plates, Faber reveals the cognitive dissonance at the heart of meat consumption.

In the chapter where Isserely visits the underground slaughter pens, the scene can be said to allude to a process of oil extraction, digging deeper to pierce the earth's crust in order to gain the desired commodity. She feels a 'humid ambience of recycled breath' (Faber 2006, 169) as if stuck down an oil well or mining for coal. Isserley is then finally confronted with the products of her labour: the vodsels' fattened and tongue-less, barely indistinguishable from one another, 'their fat little heads identical, blinking stupidly in the sudden light' (Faber 2006, 169). The uncanny resemblance between the scene of oil extraction and the slaughter pens suggests that these industrial practices are in some way connected as the corporeal boundaries of the earth become torn and severed just like the human flesh which is awaiting manufacture in the text. It is also a stark reminder of the live and dead organisms which form our fuel for both food and transportation. Inside the pens both Isserley and the reader cannot avert their attention and gaze from the horror before them: the stench of bodies fattened, with tongues burnt out, huddled together in an undistinguishable mesh. In this moment, we are drawn to what Nixon describes as a different form of witnessing, 'of sights of the unseen' (Nixon 2013, 15). It is a way of addressing the violence which is often disavowed when living beings are turned into objects to be consumed or tools to be utilised. As Amlis Vess states, his father often 'describes [the vodsels] as vegetables on legs' (Faber

2006, 167). By becoming inanimate, the 'vodsel' becomes inert and passive without resistance allowing for an unchallenged human authority over their existence. Faber's reversal of the human/animal binary remains a poignant reminder that such power relations are indeed arbitrary and foregrounded through processes of naturalisation and social construction.

Amlis Vess is shocked that the vodsels appear to have a language. He witnesses one vodsel scuttle along the ground of the pen attempting to communicate with both himself and Isserley. With the strands of hay within the slaughter pen, the vodsel conjurers the word 'MERCY' (Faber 2006, 171). Isserley racks her brain for a translation but soon discovers the word is untranslatable in her own tongue. Language is inadequate to communicate across species lines, the words are empty conveying little from one being to the other. The vision of animal suffering, however, remains and it is a sight of violence which is rarely brought into view. Such an act is one of activism, addressing and exposing what happens in peripheral and secluded spaces. According to Nixon, fiction can be a way of making the unapparent appear (Nixon 2013, 15). Faber embarks on such a task, unveiling human exceptionalism and confronting the reader with both the edibility of the human itself, as well as the unseen conditions of the factory farm. In the secluded corners of Inverness, Faber exposes us to the different forms of violence that are unleashed on those classified as nonhuman. It is to those externalised environments, those which are treated as 'out there', beyond sight and recognition to which both Faber and Okorafor turn. These externalised sites and peripheral landscapes are treated as what Nixon describes as 'a separate non-renewable resource' (Nixon 2013, 17). But as the WReC group reiterate in their interpretation of combined and uneven development, the maldevelopment of the periphery is essential to the growth of the imperial core. These spaces of extraction and consumer centres are therefore not entirely separate entities, they are bound and connected where one feeds off and oppresses the other, creating a co-dependency and maintaining systems of inequality. By turning to 'sights of the unseen' (Nixon 2013, 15), Faber and Okorafor provide a magnification of the periphery following the flow of commodities from their birth and their origin.

The human becomes a site of contestation in both *Lagoon* and *Under the Skin* where the characters are confronted with their own animality as well as the alterity of others. In the case of Faber's Isserley, she must reiterate her 'humanity' in order to continue her work for Vess Industries. In so doing, she can enforce clear demarcations between herself and her 'food'.

As she informs Amlis Vess, 'I hate to tell you this, but I really doubt there's much similarity between the way you and I live and breathe, let alone me and [...] my breakfast' (Faber 2006, 164). Her need for distinction is also constantly stressed by an emphasis on cleanliness. In her work Isserley must lure and collect lost and forgotten hitchhikers from the motorway; once in the passenger seat they are injected with a drug which leaves them unconscious. Once this is done, she disguises them and drives back to the Vess Industries processing plant for the body to be processed for meat production. In this role, she must interact and engage with those she classifies as animals. She expresses a disgust when one 'vodsel' happens to touch her as if his body contained a contagion. He touches her arm in gentle compassion, and she attempts to forget about the moment instantly. Her arm hardens at his touch and following this resolves to wash and put fresh clothes on. However, the surgery which has been conducted on her body in order for her to be deceptively recognised as a 'vodsel', contradicts her reinforcement of 'human' borders. These surgical modifications are in the interest of Vess Industries for her to carry out her role; it is a job which allowed her to escape from the impoverished estates of her own planet but a position which nevertheless challenges her notion of what it means to be human. Her essential need therefore to reassert categorical borders is of utmost importance for the preservation of her human identity. As Adams crucially notes in *The Sexual Politics of Meat*, one must view the animal before them not as a living being but inert and passive. For Isserley to carry out her work, she must do the same, but she must also have an understanding that she herself is human. This narrative of the 'human' becomes a defence mechanism and simultaneously a way of preventing communication and connection across species divisions.

For Isserley to maintain a coherent understanding of her humanity, she must consistently dehumanise the vodsels. This serves to perpetuate the power relations of this alien capitalist system. This process of dehumanisation is central to the flow of commerce and the continuation of meat production. The privileged classes from her alien planet only encounter the 'vodsels' as sanitised bloodless meat products, rarely ever apprehending what was once the living being which is now on their dinner plates. This sense of detachment and erosion of human and nonhuman animal encounters reflects our own capitalist system. As Sherryl Vint suggests in her analysis of John Berger's work, 'capitalism has irredeemably isolated man, who can no longer share an exchange of mutual looks with other species, whom he has marginalised or destroyed' (Vint 2010, 10). Through the perspective of Isserley, Faber produces what Nixon describes as a 'different form of witnessing' (Nixon 2013, 15). Her perspective as an alien species looking on to what we understand as human life serves to defamiliarise the notions of humanity and animality, revealing them as arbitrary borders. Isserley is also on the frontline of the production process, where her interaction with other life forms is driven by the external imperative of capitalist consumption. Unlike the modern consumer, Isserley is not entirely severed from encounters with nonhuman animals but must capture them as part of her position at Vess Industries.

After Amlis Vess discovers the 'vodsels' have a language of their own, Isserley reflects on this and the suffering they've endured. She rejects this fleeting empathetic thought and says to herself,

> Just look at these creatures! Their brute bulk, their stink, their look idiocy, the way the shit oozed up between their fat toes. Had she been so badly butchered, brought so close to an animal state physically, that she was losing hold of her humanity and actually identifying with animals? If she wasn't careful, she would end up living among them, cackling and mooing in meaningless abandon like the cavorting oddities on television. (Faber 2006, 172)

In this description, Isserley emphasises the 'vodsel's' corporeality, the bodily defecation, the absence of intelligence and reason. The same logic commonly used to exclude nonhuman animals from moral communities is articulated here in Isserley's reasoning for not valuing their life. Their look of 'idiocy' stresses their lack of reasoning capabilities, which is often attributed in political and moral philosophy as the defining characteristic that distinguishes the 'human' from other animals. But as Plumwood articulates, reason has often been a tool of oppression in a system of dualisms which maintain unequal power relations of gender, race, class and species.

The magnification of the hidden enclosures of the meat processing plant in *Under the Skin* and the exposure to the impacts of petroculture in *Lagoon* have opened us to Nixon's 'sights of the unseen'. There is an exploration of the origins of commodities and the peripheral worlds of production in the two texts, which has also allowed for the capitalist story of underdevelopment to become magnified. However, in order to break from a capitalist imperial ideology and to alternatively embrace new ways of being and living in the environment, it is not a projection of humanist ideals which is the solution but instead their retraction. Like Anat Pick, I call for a radical 'contraction of humanity rather than its benevolent extension to nonhumans' (Pick 2011, 414). What if we were to embrace a type of animality connected and bound to other species and the world itself? In such an approach, moving away from the human is a positive project and thus can be a way of rejecting its colonial and extractivist logic. I now go on to explore the breakdown of the humanist and colonial project while addressing the implications of an encounter with an alien/animal alterity and an earthly vulnerability within the two texts.

Exposure and Resistance: An Ethics of Vulnerability in *Lagoon* and *Under the Skin*

An attentiveness to vulnerability is central to the planetary and earthly politics and ethics I hope to engender in this book; specifically, it is the theoretical foundations of Pick and Butler's research on ethical responses to suffering which helps us move towards this form of earthly attention. The scholarship of Josephine Donavan and Carol J. Adams focuses on an ethics of care informed by feminist practice and animal ethics. Likewise, ecofeminist figures such as Plumwood promote a relational subject outside of the dualisms which segregate the human from their wider ecology. The planetary and vulnerable ethics I foreground offers these forms of fellowship and integration with other creatures. As explored earlier, Butler's later work brings ethics to the fore through an emphasis on precarity and vulnerability. In both *Precarious Life* and *Frames of War*, Butler explores an ethics and politics of precarity inspired by Emmanuel Levinas, where the vulnerability of the self and the Other is central to the encounter with alterity. What Pick, Butler and Levinas all recognise, however, is 'that the threat of violence hangs over the encounter with alterity' (Pick 2018, 416). Recognition or a shared vulnerability does not necessarily lead to Donna Haraway's notion of a kinship but rather can be an encounter pervaded by both fear and violence. Nevertheless, such an encounter with vulnerability is a mode of exposure where, as Pick suggests, there 'is a threat of violence or the possibility of care' (Pick 2018, 414). Drawing upon feminist, decolonial and animal studies scholars, I suggest that their exploration of vulnerability thus opens up a challenge to human mastery, domination and human exceptionalism. The question, however, is whether such exposure to nonhuman vulnerability can disrupt the flow of 'life' becoming commodity, the capitalist market relations which turn different species and the environment into objects of consumption. It also leads to a further question, one which asks if vulnerability can provide a mode of resistance to capital's extractive projects.

Taking vulnerability as mode of exposure as our cue, I now explore the moments in the two texts where exposure to a helpless vulnerability across species lines and to the earth itself creates a transition in action. These moments are not necessarily a guarantee for ethical change, but as Butler and Pick emphasise in their work, it can possibly be the opportunity for violence. The encounters with alien and nonhuman life forms in the texts can, in certain cases, become a wounding, a recognition of one's own flesh and penetrable borders which can lead simultaneously to an invitation of recognition or a violent self-preservation. What is most interesting, however, is how a wound which remains metaphorically open, as well as an exposure to the vulnerability of different life forms, can often question and challenge human mastery and colonial logic. In order to present a decolonial vision and promise, a world beyond capitalist forms of production, the two narratives expose a corporeal vulnerability leading to other ways of feeling and seeing the world beyond the 'domain of the human' (Pick 2011, 2).

Okorafor's text is an encounter with alterity and vulnerability of the Other, involving an amalgam of sea life, nonhuman species and alien life forms. Vint suggests one of science fiction's 'most promising' themes is the 'aspiration that humans might interact with an intelligence other than our own and be transformed by it' (Vint 2010, 4). Simultaneously, this interaction with an animal/alien Other can be an opportunity of violence, one which illuminates not only transversal relations with the nonhuman but also relations of exploitation and oppression. Okorafor presents these two poles in her narrative exploring acts and events of violence while pointing towards an opportunity of ethical recognition across species lines.

The discovery of corporeal vulnerability in Okorafor's *Lagoon* comes with the revelation that the alien Ayodele can experience pain. As previously noted by Plumwood, invulnerability is the coloniser's logic, an exposure to vulnerability, however, is perhaps its undoing. In the chapter 'Greetings', Ayodele announces her arrival to the people of Lagos. She declares, 'we are guests who wish to become citizens … here. We chose here. I am the first to come and greet you' (Okorafor 2016, 111). This greeting, however, is unwelcome—Ayodele is perceived as a threat as an officer yell 'Shoot it! […] kill it! Kill it!' (Okorafor 2016, 136). Ayodele is identified as an 'it' separate from a moral community, a phantom which must be expelled. As Butler notes, 'the human is not a universal condition, it is a category, and we decide who counts as part of it and who doesn't' (Butler 2012, 20). In this case, Ayodele and her people are considered a threat as they deviate from the normative frame of the human. If, as Butler contends, violence, mourning and loss can be the basis for a political community, how does the violence wrought upon Ayodele change the course of action in the narrative?

Ayodele's body is brutally attacked by the soldiers: 'they shot her in the thighs, chest, and face, everywhere' (Okorafor 2016, 136). Adaora describes 'her fragile, greying body [...] hopping and jerking on the ground' (Okorafor 2016, 136). The emphasis on her fragility exposes the vulnerability of the alien body as akin to our own, a likeness within a shared condition of suffering. Adaora contemplates that 'maybe Ayodele responded so strongly because they made her experience pain [...] the way she was screaming and thrashing, she was not just in pain, she was shocked to be in it' (Okorafor 2016, 140). This discovery of vulnerability of another and the subjection to violence is a reminder of a co-dependency, where one can be affected and hurt by others. After this experience and shock of being made vulnerable and in pain, Ayodele uses her shapeshifting abilities to transform herself into a monkey. It is particularly significant that she chooses to take the form of a nonhuman animal after such an experience. The movement suggests how vulnerability can connect us to other species. Invulnerability is unravelled in such an experience where, like Plumwood's account of 'becoming prey', there is reorganisation of hierarchical relations.

This cohabitation of alien life forms, human beings and animals in *Lagoon* illuminates the violence wrought upon those who deviate from the normative understanding of the 'human'. As Butler argues,

> It seems to me that even in situations of antagonistic and unchosen modes of cohabitation, certain ethical obligations emerge. First, since we do not choose with whom to cohabit the earth, we have to honor those obligations to preserve the lives of those we may not love, those we may never love, do not know, and did not choose. (Butler 2012, 150)

These ethical obligations may not always emerge, as there are those who may respond to difference with modes of oppression and violence. In *Lagoon* we witness the response to alterity as both an invitation for 'violence or the possibility of care' (Pick 2018, 414). Encounters with nonhumans can illuminate a recognition of difference and oppositions, which simultaneously may lead to a shoring up of borders, a violent form of self-preservation, or alternatively an openness and a hospitality to the Other.

Okorafor concludes the narrative from the perspective of a spider; the spider announces, 'I am the unseen' (Okorafor 2016, 291). The spider has been there for centuries spinning their web and tapestry of the tales of Lagos. The spider declares it 'feels the press of other stories' (Okorafor 2016, 292), where it opens its web to the peripheral narratives and lives, incorporating them within a wider web of events and experiences. Okorafor open us up to the 'unseen' (Okorafor 2016, 291), those who are edited out of what is considered normatively human, exposing a corporeal vulnerability which moves across species lines towards the alien and the Other.

The science fiction trope of an alien invasion not only resonates with colonial legacies and its history but also allows one to frame our relationship to nonhumans. Okorafor's text presents alien shapeshifters, insects, marine life and other species, binding the reader and the characters to them in usual ways. There is a strange and beautiful bond formed through vulnerability and a way of 'reconnecting with our embodied being, what might be thought as our animal nature' (Vint 2010, 9). Vint notes a resonance between animal studies and science fiction, the ways they both present an encounter with the strange and unfamiliar:

> Both take seriously the question of what it means to communicate with a being who's embodied, communicative, emotional, and cultural life – perhaps even physical environment – is radically different from our own. (Vint 2010, 1)

Science fiction allows for the event of an ethical moment or conversely a perhaps violent interaction with an Other's vulnerability who is radically different from ourselves. The genre's terrain of foreign lands, alien bodies and strange encounters enables a process where the human becomes challenged from their position of mastery in which it could lead to an onslaught of a violent attack or alternatively a plea and recognition of vulnerable life.

Michel Faber's *Under the Skin* likewise transforms and interrogates the 'human' through the logic of vulnerability as well as by acts of violence which expose the construction of the human as fragile, existing within a wider set of relations and ecology. Isserley's human exceptionalism becomes undone in the face of violence and a confrontation with the other. For Sherryl Vint, SF confronts us with 'the gaze of "absolute alterity", an Other who looks back at us from its own point of view and often one whom we must acknowledge as having power comparable if not identical to our own' (Vint 2010, 11). It is this experience of becoming confronted with power and violence, which effects Isserley's cold and managerial form of detachment towards the vodsels as nothing but 'mere' animals.

The unravelling of Isserley's human exceptionalism comes at severe cost when she endures an incident of sexual violence from a hitchhiker she collects from the road. As he forces her to obey his commands, she desperately attempts to search for the right word as plea for the perpetrator to stop. Her mind returns to the incident with Amlis Vess and the vodsel who writes mercy upon the ground to prevent his slaughter. She had never said the word before and pleads 'Murky' (Faber 2006, 186). The words fail to communicate her suffering and her fear. Repeating the vodsel's words, she becomes aligned with the animal, an absolute and bare vulnerability to a force beyond her control.

This incident breaks down her cold detachment to the vodsels, and instead, it is replaced by a need for a bitter revenge. Isserley takes her revenge by heading down once again to the slaughter pens back at the meat processing plant at Albach Farm. She comes to watch the slaughter of one of the vodsels to witness their pain and suffering. As the workers slit one of the vodsel arteries, she involuntarily screams 'Yes!' (Faber 2006, 219) satisfied by the violence happening before her. The workers are stunned by her emotional response where one of them turns to Isserley and states: 'I'm sorry, Isserley [...] but I don't think it's a good idea for you to be here [...] we are doing a job here [...] feelings don't enter it' (Faber 2006, 219). Isserley's trauma leads to a rupture of the managerial approach of capitalist imperialism. She is removed from the premises because in the act of slaughter, killing animal life is merely a job—one of cold detachment where the being before you is already an object. Isserley's experience of vulnerability is a severe recognition that one is not entirely independent and in control; the human's assumption of invulnerability is informed by this logic of colonisation which unravels in the face of a violent attack.

Through this exposure of bodily and earthly sense of vulnerability, Isserley is not only provoked into a desire for revenge but also into acts of care. In the final stages of the novel, she picks up a hitchhiker but not for the purposes of Vess Industries. The hitchhiker she picks up is on his way to his pregnant partner and in desperate search for a lift. She has no intention of taking him to Albach Farm for the slaughter, but rather just as an act of hospitality. The conversation between the two of them veers into the afterlife and what becomes of their bodies once they have left them; the hitcher retorts, 'Who knows, eh? Ah might come back as a wumman, or a wee beastie!' (Faber 2006, 292). This response reiterates the central theme of crossing species division, of opening oneself up to another, of embracing an alien animality and difference.

In the end, Isserley imagines her own body an array of 'atoms' mingling 'with the oxygen and nitrogen in the air' (Faber 2006, 296). She envisions her death where 'her invisible remains would combine, over time, with all the wonders under the sun. When it snowed, she would be part of it, falling softy to earth, rising up again with the snow's evaporation' (Faber 2006, 296). Her imagined death captures her as becoming one with the earth. With a trembling hand she declares 'here I come' (Faber 2006, 296) towards her demise. The human in the text is no longer a monolithic and violent force but becomes merged with the elements of the earth. Isserley's exposure to vulnerability creates a rupture within the successive flow of meat resources; she turns away from her role and her position at Vess Industries to instead merge and combine with the wider ecology of the planet. This rupture, however, is perhaps a mere inconvenience for this capitalist system. They will replace Isserley with another worker and the production will perhaps continue unabated with business continuing on as usual.

Exposure to earthly vulnerability ignites individual transitions within the two narratives leading to both acts of violence and ones of care. Its challenge to human mastery becomes evident as the characters endure an unbearable vulnerability which reveals a sense of co-dependency, a reliance on the other before them whether it be human, animal or alien. As Pick suggests, 'universal and shared, vulnerability blurs species distinctions because humans and nonhumans alike are subject to natural law, to injury, and to death' (Pick 2011, 411). This revelation of the human's fleshy borders and bodies which become edible unravels the humanist project that is driven by a colonial extractivism. Pick's notion of creatureliness, which is 'the state of being exposed to natural necessity and the ravages of power' (Pick 2011, 414), is revealed to the characters in varying ways through invasion, assault, love and care. In the face of the continuation of mass meat production and energy regimes, however, the creaturely and an ethics of vulnerability demonstrate certain limitations within the texts.

The affective change that is manifested through an exposure to vulnerability of the self and the Other serves to ignite individual transitions amongst the characters, but the capitalist system of production remains in flow. In *Lagoon* a techno-utopianism emerges and is embraced, but presumably the capitalist system of exploitation within energy regimes remains. Likewise, Isserley herself may have escaped the exploitative labour of the meat industry yet Vess Industries' meat plantation project continues to grow. The ruptures and breaks created in the texts are thus reorganised back into the continuous machine of energy and meat production, as capitalism continually renews itself, evoking disasters and moving through systematic erasures—occluding the history and the lives of those who fall outside the category of the human.

Nevertheless, acts of resistance, however small and minute, are found in these narratives as they challenge the present violence of energy production, discovering ways to resist its exploitation. Okorafor and Faber explore how violence and care can occur in tandem, where hopelessness and resistance emerge alongside one another. In Jose Esteban Munoz speech 'the brown commons', we find a similar call for fight and hope in the most severe and treacherous of conditions. As he reads, 'It was Leon Trotsky who famously said, active indignation is linked with hope, this hope is what I've called an educated hope or a critical desire. It is an affective refusal and salient demand for something else' (Muñoz 2020, 205). *Under the Skin* and *Lagoon* often refused the grammars of the extractive zone, demanding alternatives to extractive capitalism, while attentive to the hope which the discovery of earthly vulnerability provides. Their fight for change may not yet have entirely succeeded but the promise of hope persists; it demands 'the creation of emergent alternatives' (Gómez-Barris 2017, 4).

References

Adams, Carol J. 2020. *The Sexual Politics of Meat: A Feminist-Vegetarian Critical Theory*. Twentieth anniversary edition. Bloomsbury Revelations Series. London: Bloomsbury Academic, An imprint of Bloomsbury Publishing Inc. https://doi.org/10.5040/9781501312861.

Alessandrini, Anthony C. 2009. The Humanism Effect: Fanon, Foucault, and Ethics without Subjects. *Foucault Studies*, September, 64–80. https://doi.org/10.22439/fs.v0i7.2637.

Anderson, Kay. 2012. *Race and the Crisis of Humanism*. Taylor and Francis.

Butler, Judith. 2012. Precarious Life, Vulnerability, and the Ethics of Cohabitation. *The Journal of Speculative Philosophy* 26 (2): 134–151. https://doi.org/10.5325/jspecphil.26.2.0134.
Callaghan, James. 2015. Oil: A Crude History of Britain. https://www.bbc.co.uk/mediacentre/proginfo/2015/36/oil-a-crude-history-of-britain.
Faber, Michel. 2006. *Under the Skin.* San Diego [etc.]: Harcourt.
Gómez-Barris, Macarena. 2017. *The Extractive Zone: Social Ecologies and Decolonial Perspectives.* Dissident Acts. Durham; London: Duke University Press.
Harootunian, Harry. 2000. *History's Disquiet: Modernity, Cultural Practice, and the Question of Everyday Life.* Columbia University Press.
Harvie, Christopher. 1994. *Fool's Gold: The Story of North Sea Oil.* 1. publ. London: Hamilton.
Helman, Christopher. n.d. Swordfish Attacks BP Oilfield In Angola. *Forbes.* www.forbes.com/sites/christopherhelman/2014/03/14/swordfish-attacks-bp-oilfield-in-angola.
Lazarus, Neil. 1999. Disavowing Decolonization: Fanon, Nationalism, and the Question of Representation in Postcolonial Theory. In *Frantz Fanon: Critical Perspectives.* Routledge.
Maass, P. 2010. *The Violent Twilight of Oil.* London: Penguin Books Ltd: [distributor] Penguin Books Ltd.
McKittrick, Katherine. 2015. *Sylvia Wynter: On Being Human as Praxis.* Duke University Press.
Muñoz, José Esteban. 2020. *The Sense of Brown: Ethnicity, Affect and Performance.* Edited by Joshua Takano Chambers-Letson and Tavia Nyong'o. Perverse Modernities. Durham: Duke University Press.
Nixon, Rob. 2013. *Slow Violence and the Environmentalism of the Poor.* First Harvard University Press paperback edition. Cambridge, MA; London: Harvard University Press.
O'Connell, Hugh Charles. 2016. 'We Are Change': The Novum as Event in Nnedi Okorafor's *Lagoon. The Cambridge Journal of Postcolonial Literary Inquiry* 3 (3): 291–312. https://doi.org/10.1017/pli.2016.24.
Okorafor, Nnedi. 2016. *Lagoon.* Saga Press.
Paye, Michael. 2019. Beyond a Capitalist Atlantic: Fish, Fuel, and the Collapse of Cheap Nature in Ireland, Newfoundland, and Nigeria. *Irish University Review* 49 (1): 117–134. https://doi.org/10.3366/iur.2019.0384.
Pick, Anat. 2011. *Creaturely Poetics: Animality and Vulnerability in Literature and Film.* New York: Columbia University Press.
Pick, Anat. 2018. Vulnerability. In *Critical Terms for Animal Studies.* Chicago University Press.
Plumwood, Val. 1995. *Human Vulnerability and The Experience of Being Prey.* Quadrant Magazine Ltd.

Pudnt, Johanna. 2018. Organic Fantasy and the Alien Archetype in Nnedi Okorafor's Lagoon" *Exploring the Fantastic*, no. Verlag.
Shukin, Nicole. 2009. *Animal Capital: Rendering Life in Biopolitical Times.* Posthumanities 6. Minneapolis: University of Minnesota Press.
Singh, Julietta. 2018. *Unthinking Mastery: Dehumanism and Decolonial Entanglements.* Durham; London: Duke University Press.
Szeman, Imre. 2014. Blind Faith. *Policy Options.* https://policyoptions.irpp.org/magazines/technology/blind-faith/.
Szeman, Imre. 2015. Entrepreneurship as the New Common Sense. *South Atlantic Quarterly* 114 (3): 471–490. https://doi.org/10.1215/00382876-3130701.
Vint, Sherryl. 2010. *Animal Alterity: Science Fiction and the Question of the Animal.* 1. publ. Liverpool Science Fiction Texts and Studies 39. Liverpool: Liverpool Univ. Press.
Warwick Research Collective. 2015. *Combined and Uneven Development: Towards a New Theory of World-Literature.* Postcolonialism across the Disciplines 17. Liverpool: Liverpool University Press.
Watts, Michael. 2014. Oil Frontiers: The Niger Delta and the Gulf of Mexico. In *Oil Culture.* University of Minnesota Press.
Weil, Simone. 2021. *Simone Weil: An Anthology.* Penguin Classics.
Wenzel, Jennifer. 2006. Petro-Magic-Realism: Toward a Political Ecology of Nigerian Literature. *Postcolonial Studies* 9 (4): 449–464. https://doi.org/10.1080/13688790600993263.
Wilson, Sheena. 2012. Sight, Site, Cite: Oil in the Field of Vision. *Imaginations: Journal of Cross-Cultural Image Studies* 3 (2): 4–5. https://doi.org/10.17742/IMAGE.sightoil.3-2.1.
Wood, Ellen Meiksin. 2005. *Empire of Capital.* Verso.
Wynter, Sylvia. 2003. Unsettling the Coloniality of Being/Power/Truth/Freedom: Towards the Human, After Man, Its Overrepresentation--An Argument. *CR: The New Centennial Review* 3 (3): 257–337. https://doi.org/10.1353/ncr.2004.0015.
Yusoff, Kathryn. 2018. *A Billion Black Anthropocenes or None.* Forerunners Ideas First from the University of Minnesota Press. Minneapolis, MN: University of Minnesota Press.

CHAPTER 4

Energy Infrastructure: Oil Spills and Roadkill in Peripheral Literatures and Visual Cultures

In their writings, the Out of the Woods Collective argue that grief and mourning can be an immobilising component when grasping the cascading scales of the destruction of climate change. As they suggest, 'the realisation that people, creatures, and entire ecosystems have died, and will continue to die does not immediately lead to determination, but melancholia' (Out of the Woods Collective 2020, 2). This chapter draws on such experiences of melancholia, grief, despondency, pain and disgust in response to nonhuman death and environmental destruction which is a consequence of energy infrastructure and automobility. I suggest, however, how these affects can lead to political and ethical calls for resistance against nonhuman suffering and ecological crisis. For the Out of the Woods Collective, ecological damage and the death of other species are about attuning our senses to the microscopic, as well as the grand scales of destruction. As they argue, 'to really grapple with the scale of destruction involves attending to the slower, less eye catching processes: the pollution and erosion of the soil; the feeling of forests that bind them together; the extinctions of creatures that feed on and were fed by the Earth' (Out of the Woods Collective 2020, 2). A recurring question in this book is how do we register suffering? How do we acknowledge the vulnerability of nonhuman life forms—what forestalls such ethical recognition and thus acts as a blockade to political and ethical change? I thus turn to the of role affects in order to examine such questions—in the hopes of refusing to

J. Taylor, *The Nonhuman Narratives of Energy*, Palgrave Studies in Animals and Literature,
https://doi.org/10.1007/978-3-031-94332-4_4

turn away from what should be acknowledged and perceived. I argue that the affective registers in scenes of environmental ruin and violence open up the possibility of recognition and point towards Anat Pick's *Creaturely Poetics*. Pick's emphasis on the communicative pressures of vulnerability, of 'tracing the logic of flesh in examples across image and text' (Pick 2011, 6), provides an important nexus in my understanding of affect and its ethical and political potential in the chosen sources. Reading grief, disgust and pain under Pick's creaturely prism poses a mode of non-anthropocentric feeling, an encounter with the 'flesh and blood vulnerability of beings' (Pick 2011, 3). At its core it is a reminder of what interdependency and a vulnerable politics and ethics leads us to.

Exploring extractive zones, edges of road infrastructure and contaminated waterways, I look to the affects felt in response to scenes of animal death and environmental devastation; the multitude of emotions from neglect, alarm, grief, disgust and hope, inviting questions of how we act and why we act. This chapter explores the role of affects in igniting ethical and political action exploring the work of Judith Butler, Sianne Ngai, Sara Ahmed and Anat Pick. I open the chapter with this theoretical exploration of affect and the nonhuman life, examining how acts of mourning, reactions of disgust and what Ahmed terms the 'sociality of pain' can invite a response to environmental ruin and the death of different species.

Having set out my theorisation on affect, I then move to analyse Steve Baker's exhibition *Roadside* focusing particularly on the role of Butler's notion of grievability in relation to the images of roadkill while also analysing sensations of disgust in response to the rotting animal body. I further address how affect is structured and mediated in the images and explore the ways in which collecting images of animal remains is a work of memorialisation, gesturing towards a non-anthropocentric or creaturely mourning. Moving to another form of violence and ruin, I turn to the oil spills and flares captured in Helon Habila's novel *Oil on Water*. I introduce the Nigerian literary context and its legacy of environmental writing while contextualising the colonial and neo-colonial extraction history. My specific focus in Habila's text is the devastation of the Niger Delta landscape. I argue that feelings of pain, disgust and despair in response to this devastation underpin a movement towards environmental and political action against violent energy regimes. Following this focus on environmental resistance, I analyse Rita Wong and Cindy Mochizuki's *Perpetual* opening with Wong's experience of the Indigenous protest healing walks in the Tar Sands. By addressing the diverse affective responses of grief, despair and hope in the face of violent and neo-colonial energy practices within

Perpetual, I argue that such affective encounters are central to forming a creaturely and non-anthropocentric relation to energy, a recognition and fellowship with other life forms which moves away from violent practices of extraction.

Theorising Affect Beyond the Human

Grief is often regarded as a solitary experience and situated primarily within the domain of the human and where the bodies that are grieved tend to fall into this category. Mourning, however, can be a collective practice, one of sociality rather than something privatising and can also include nonhuman communities and the earth itself. I want to suggest then, in line with Butler's argument in *Precarious Life*, that grief can be political, furnishing recognition of the lives and deaths that go unnoticed.

In Freud's *Mourning and Melancholia*, he asks 'what is the work that mourning performs?' (Freud and Whiteside 2005, 204). I also ask this question but in its relation to nonhuman life forms, the response to environmental disasters and nonhuman death considering the implications of mourning subjects that are often deemed ungrievable. In Freud's description of the work of mourning he suggests,

> Serious mourning, the reaction to the loss of a loved one, contains the same powerful mood, the loss of interest in the outside world – except as it recalls the deceased – the loss of ability to choose any new love object – which would mean replacing the mourned one – and turning away from any task that is not related to the memory of the deceased. (Freud and Whiteside 2005, 204)

There is a sense of insularity in Freud's description, an act of turning away from the world, rejecting other social bonds and interdependency for a solitary existence, a life shaped by an attachment that is now severed. However, like Butler, I look to the 'relational obligations' (Butler 2020, 64) that can emerge from the act of mourning, the sense of worldliness which can arise from caring, loving and mourning another. Mourning in this context thus becomes a method of fashioning a political community, highlighting sociality and fellowship with those who are within and beyond our sensory field.

What does it mean to live in a world in which grief is distributed equally, the idea of what Butler terms 'equal grievability' (Butler 2020, 61)? Would the violence wrought through extractive regimes move from the periphery to the centre stage? What resides in the margins, excluded from the realm of perceptibility, therefore becomes an important space of attention. In the face of increasing violence against nonhuman life forms, it is vital to ask whether this someone or something ever counted, if in fact it was ever considered living, would its life register as a loss, a life worthy of mourning? For Butler, the exclusionary mechanisms of grief and mourning are bound within the logic and construction of the human: 'the field of the human is constituted through basic exclusions, haunted by those figures that do not count in its tally' (Butler 2020, 59). The 'human' is not a neutral category but is rather connected to colonial imperatives and regimes, haunted by legacies of power and immersed in histories of violence. The label of the human, in some cases, entails a degree of protection that a life is worth saving and a death worth mourning. What of those who do not belong or who are cast out of this category, are they too considered worthy of grief and attention? Is it instead possible to conceive of a non-anthropocentric mourning? Like James Stanescu's revision of Butler's work in his article *Species Trouble* (Stanescu 2012) I too propose a non-anthropocentric relation to mourning where the 'human' is not the guiding model of who and what is considered grievable. The disavowal, deflection and repression of grief are central to the experience of mourning nonhuman life forms—how do we openly mourn those never considered a life in the first place and in doing so turn our grief into social and environmental movements for the future?

Butler suggests that certain lives become recognised as worthy of attention, care and mourning through the organisation and structuring of what she terms the 'frame' that is dictated around schemes of recognisability. The interpretative schemas of recognisability foreground what lives and what events are worthy of regard and attention. As Butler asks, how can these frames of recognisability 'help us understand why it is that we might feel horror in the face of certain losses but indifference or even righteousness in light of others' (Butler 2016, 42). The frame is thus an essential mediation of affective responses, designating who and what is recognised as a grievable subject. The 'being' of life itself, as Butler argues, is constituted through selective means (Butler 2016, 1). To rupture the exclusionary mechanisms of the frame, what is left outside of its borders, I instead introduce a form of reframing through literary and visual media which gather the leftovers and remnants of violent practices. My intention is to highlight how the three chosen cultural sources provide this essential reframing, moving from the margins to the centre, foregrounding nonhuman death and the devastating effects of energy culture.

Grief has the potential to forge connections across species lines while exposing which lives are framed as grievable subjects. Sensations of disgust, however, can also point towards proximity and connection beyond the domain of the human. Disgust, for Sara Ahmed, is a 'contact zone' (Ahmed 2004, 87) forcing a proximity to what can often be ignored or rejected. Introducing Sianne Ngai's reading of the political potential of disgust, alongside Ahmed's analysis of disgust, I demonstrate how disgust can lead to different avenues of fellowship with nonhuman life forms. As Danielle Sands has argued, 'empathy and disgust share unexpected similarities: while perceived as humanising or even civilising, they both draw upon "the common animality" shared by humans and other species' (Sands 2022, 173). My exploration of disgust facilitates this common animality and proximity to nonhuman life forms.

Crucial to the affective experience of disgust is the element of paradox, both a bodily intrusion and an expulsion, a fascination alongside repulsion. In Winfried Menninghaus' work, *Disgust: Theory and History of a Strong Sensation*, he suggests that disgust engenders both 'intolerable contact and a union' (Menninghaus 2003, 2). Disgust becomes at once an invitation for the possibility of attention, a moment to linger, to encounter or alternatively a refusal of recognition and a physical rejection. By exploring disgust and the responses it provokes, I ask what ethical demands does disgust call for in our interaction with nonhuman life? In Elenora Joensuu's *Politics of Disgust*, she asks similar questions around the relationality and ethical dimensions of the disgust encounter. She invites the questions: 'what others have you encountered in your experiences of disgust? If you were to re-examine a previous disgust encounter, what others—real or imagined—does the encounter evoke? Who is your disgust in relation with?' (Joensuu 2019, 118). This re-examination of disgust to centre on questions of the interaction with another invites what Joensuu describes as 'the possibility of connecting to experiences beyond our own' (Joensuu 2019, 121). Disgust thus becomes a gateway to an encounter, perhaps even an empathetic engagement with another. Its dual nature, however, of both recognition and dismissal provokes important questions around how we interact with the nonhuman subjects that intrude into our sensory field.

The sense of rejection that occurs within disgust brings us to Pick's discussion of exposure and deflection. Pick quotes Cora Diamond when referring to this complex duality; Diamond suggests that deflection 'happens when we are moved from the appreciation, or attempt at appreciation, of a difficulty of reality to a philosophical moral problem in the vicinity' (Diamond 2003, 57). In the encounter with another's experience, there is both an invitation to a creaturely fellowship or the process of deflection when exposed to bodily vulnerability. We find the same dual dimensions with disgust, as an exposure or a bodily intrusion poses the possibility of either recognition or deflection. In my analysis of disgust, I consider the implications of refusing to deflect but rather to remain with the sensation which generates disgust.

Sianne Ngai's project in *Ugly Feelings* is one of recuperating disgust's animalities, generating what she describes as a bestiary of affects: 'in other words, it is one filled with rat and possums rather than lions, its categories feeling generally being, well, weaker and nastier' (Ngai 2009, 7). By introducing Ngai's non-cathartic and often amoral affects, we focus not only on the altruistic emotions and affects but also on moments and sensations which can lead to deflection and repression, the experience of disgust and aversion. Ngai aims 'to recuperate several of these negative affects for their critical productivity' (Ngai 2009, 3). Through exploring disgust, an affect which can cause one to avert one's gaze, I suggest that like Ngai 'the poetics of disgust draws us closer to the domain of political theory' (Ngai 2009, 354) but also that of ethics. As Joensuu suggests, 'expanding our understanding of disgust is therefore an ethical question and demand' (Joensuu 2019, 9). Disgust points towards possible movements of action and resistance in the domains of politics and ethics, a way of attuning to the injustices and violence against nonhumans and the environment. Like mourning, disgust is a response which can be deeply political and structured according to anthropocentrism. The symbolic charge of animal imagery can often be fused with disgust. However, rather than simply averting one's gaze from disgust, it is worth maintaining and working with this feeling, exploring its dimensions, of why one responds in such a bodily way. As with the uneven distribution of grief, what can disgust similarly inform us about the ways power functions, how nonhuman life is regarded and framed? Ngai argues that 'disgust does not so much as solve the dilemma of social powerlessness as diagnose it powerfully' (Ngai 2009, 353). As Ngai suggests, exploring the sensation of disgust does not necessarily lead to an ethical solution yet it can remain a powerful tool of diagnosis. For my purposes, analysing disgust enables a way of magnifying the response and possible rejection of nonhuman life—when one refuses to acknowledge or register suffering.

Allowing a moment of reflection on the disgust encounter reveals how it is not simply 'gut feeling', instinctual and ingrained, but is something which is mediated and structured according to normative frames. For Ahmed, 'if disgust is about gut feelings, then our relation to our guts is not direct, but is mediated by ideas that already implicated in the very impressions we make of others and the way those impressions surface as bodies' (Ahmed 2004, 83). Ahmed's aim is to suggest how disgust is in some sense already constructed, the reaction is not unmediated or purely instinctual but rather dictated by certain histories and associations. In the case of the rotting animal corpse in Steve Baker's work, it is already framed as something that evokes the response of disgust. As Menninghaus suggests,

> The decaying corpse is therefore not only one among many other foul smelling and disfigured objects of disgust. Rather, it is the emblem of the menace that, in the case of disgust, meets with such a decisive defence, as measured by its extremely potent register on the scale of unpleasureable affects. Every book about disgust is not least a book about the rotting corpse. (Menninghaus 2003, 1)

The supposedly gut response of disgust is mediated through a specific framing which is made to repel and recoil. As an emblem of the disgust encounter, 'the rotting corpse' holds a particular symbolic charge beyond its physical impact. Ahmed refers to Butler's notion of performativity in relation to disgust, noting how the reiteration of the speech acts: 'That is disgusting!' serves to generate a set of specific effects. What this illuminates is how the affective encounter is not an entirely instinctual phenomenon but is rather a complex interaction which is filtered and negotiated according to established norms and histories.

Both grief and disgust are affects which involve contact and proximity, but how, in Ahmed's words, do 'I come to feel that which I cannot know?' (Ahmed 2004, 30). Grief and disgust are forceful encounters which bring us closer to that which I cannot know: the suffering and pain of another. This brings us to the final affect that I will now explore which is pain and specifically Ahmed's concept of 'the contingency of pain' (Ahmed 2004, 27). The contingency of pain refers to becoming a witness of another's pain, of becoming 'open to being affected by that which one cannot know or feel' (Ahmed 2004, 30).

Although Ahmed's concept of the 'sociality/contingency of pain' refers to human experience, the event of being drawn into a time and space one has not inhabited before, connecting to the pain of another, it has important implications for conveying nonhuman suffering and environmental injustice. The act of being 'moved by what does not belong to me' (Ahmed 2004, 31) brings us closer to a creaturely recognition of other life forms. For Ahmed,

> The impossibility of feeling the pain of others does not mean that the pain is simply theirs, or that their pain has nothing to do with me. I want to suggest here, cautiously, and tentatively, that an ethics of responding to pain involves being open to being affected by that which one cannot know or feel. (Ahmed 2004, 30)

What Ahmed suggests is that it is not so much about feeling the pain of another but an experience of bearing witness, being open to be affected by another's pain. Ahmed uses the example of her mother's endurance of multiple sclerosis and where her mother's pleas were not always a call for action or help but rather to 'bear witness, to recognise her pain' (Ahmed 2004, 29). The act of bearing witness has an ethical demand which declares 'I must act about that which I cannot know' (Ahmed 2004, 31). As we turn to the textual and visual sources, they present encounters with often unknown suffering and destruction. They ask us to bear witness to the pain which maybe in sight but is not fully attended to or perceived. It results in both the widening of attention and an act of exposure which are crucial experiences in the path to a creaturely fellowship.

In the following engagement with the visual and textual affects found in the work of Steve Baker, Helon Habila, Rita Wong and Cindy Mochizuki, I argue that the experiences of grief, disgust and pain provide a powerful antidote to anthropocentrism by enabling a mode of address to nonhuman suffering and devastation. Pick's question in the conclusion of *Creaturely Poetics* asks 'what are the limits of attention?' (Pick 2011, 193); it is this question which provides the foundational arc in my reading of these affects, exploring how to affect or to be affected is about being open to different forms of attention.

Roadkill Encounters with Steve Baker on Norfolk Country Lanes

Steve Baker's photographic exhibition, 'Roadside', was displayed in Germany, Australia and the UK and discussed in detail in the journal *Tierstudin* by Julia Schlosser. The photographs record the animal bodies killed by collisions with automobiles, presenting traces of bodily remains and combining them with images of the surrounding landscape and artworks. For Elspeth H. Brown and Thy Phu, 'photography excites a spectrum of feelings: faced with a violent image, you may respond with both horror and pity' (Brown and Phu 2014, 1). This spectrum of feelings is ignited by the spectator's interaction with Baker's photographs. Paying particular attention to grief, disgust and the contingency of pain, I argue that Baker's artistic practice introduces a non-anthropocentric mourning and a sense of creaturely recognition by reframing roadkill as something which should be attended to and addressed. Matthew Calarco's recent work takes the subject of roadkill seriously, reflecting on what he terms as the relationship between hyper-automobility and anthropocentrism. I likewise turn to a dialogue between anthropocentrism and energy infrastructure (modes of mobility). My exploration of Baker's work can be aligned with Calarco's project of an in-depth practice of looking and perceiving. As Calarco illustrates,

> At some point, though, I began to look. I began to look, not in the sense of simply staring at roadkilled corpses (although some of that practice was involved) but more in the sense of trying to grasp the broader significance of roadkilled animals – their lives and deaths, their interrupted relations and the deeper structures that made such deaths both so common and so insignificant to dominant culture. (Calarco 2023, 12)

In my analysis of Baker's photography, I focus on Butler's notion of the 'frame' and how this serves as a powerful tool of delineation, a way of looking without refusing recognition. By recognising what gets left out of the frame or how the frame is extended or disrupted, we begin to recognise how photography mediates and structures certain responses, and why in Butler's terms, this leads to certain lives being mourned over others. Like Jane C. Desmond, I look to 'what it may take to move these roadkilled bodies from the status of unmourned to mourned' (Desmond 2016, 141). In contrast, however, I focus not solely on the transformation of roadkill becoming a mourned subject but why we also may feel revulsion and disgust. By drawing connections between grief, disgust and pain, I demonstrate the contradictory processes of rejection and address, a movement towards fellowship or to simply feign recognition.

Describing his artistic work at the roadside, Baker emphasises the experience and 'seriousness of looking'. It is this act of observation, an enactment of an address that is visualised in Baker's montage of images—a process that aligns with Susan Sontag's theorisation of photography in which 'photographs do not explain, they acknowledge' (Sontag 2001, 111). One consistent framing technique that Baker employs is the presence of his bike intruding into the frame. He describes this decision as a 'make-shift framing device making each image as an actual encounter with a particular creatures lost life' (Baker 2012, Roadside). What Baker is doing is creating a new visual code to perceive roadkill, unravelling the very term itself by presenting it not as a sidelined or neglected image but a forward and highly visible encounter. As Sontag suggests, 'Photographs enlarge our notions of what is worth looking at and what we have the right to observe' (Sontag 2001, 3). For Sontag, the project of collecting photographs is to collect the world (Sontag 2001, 3); Baker's project, however, is to collect the world which has been left behind. As Calarco suggests, 'Roadkill is a recurrent but often unthought feature of modern life' (Calarco 2023, 1). In Baker's project of contending with the seriousness of looking, he engages with a creaturely practice concerned not primarily with the rights of animals but 'about the conditions that affect the modes of our attentiveness to them' (Pick 2011, 193). Like Pick, Baker's work engages with limits of attention, opening and widening our senses to perceive the vulnerability of the nonhuman. As Pick suggests, creatureliness is also an artistic practice: 'creaturely vulnerability sets into motion different modes of artistic expression and, crucially, an alternative poetics and critical practice rooted in what I have described as the contraction of humanism and an exploration of affliction' (Pick 2018, 414). It is through Baker's reframing of roadkill which enables different forms of attention, exposing subjects of affliction, while engaging the spectator in the revelation of their own vulnerability and animality (Fig. 4.1).

The first image I would like to examine, shown above, draws us into this spectral absence of the creature who was once a living being. It is difficult to discern the taxonomy of the species, there is fur and blood splayed across the roadside, elements of bone just about visible. This ghostly stain of the dead captures a life which in Butler's terms would have never been

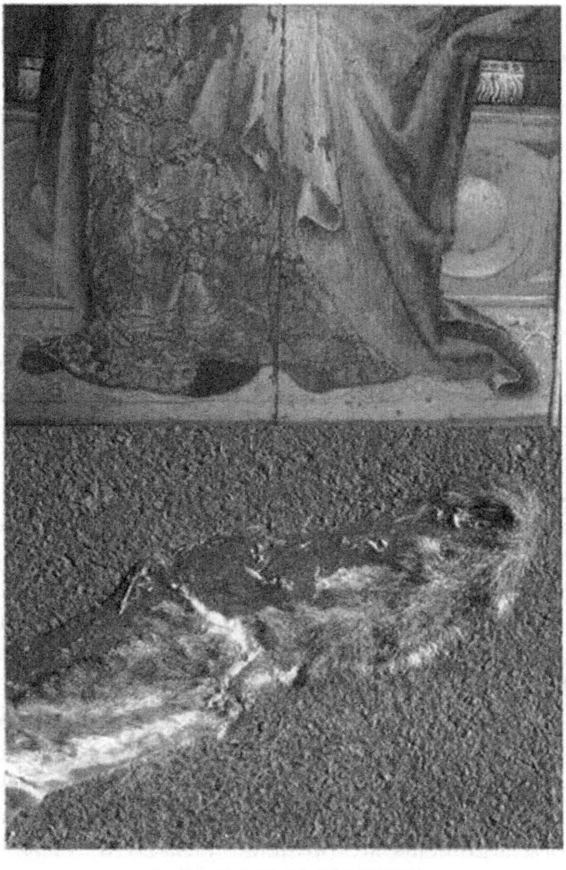

Fig. 4.1 Dobbls Lane, 22/9/09 Steve Baker

considered living, a life that in many cases fails to be captured as grievable. Baker's image alternatively draws the viewer to the traces of violence and suffering endured, magnifying the deaths which go unnoticed. John Berger suggests that 'unlike any other visual image, a photograph is not a rendering or an imitation or an interpretation of its subject, but actually a trace of it' (Berger 1991, 54). Baker thus opens us to the traces of animal remains and lives who are at the mercy of a world dominated by the automobile, an ever-expanding infrastructure orientated around oil culture. Baker centres the subject of his work on the death of the nonhuman providing a new and alternative frame to what is typically edited out. As Pick suggests, 'the editorializing that removes animals from the epistemic, legal, and emotional frameworks that would make their lives matter ensures that violence continues and animals go ungrieved' (Pick 2018, 415). Baker's photography works against this editorial removal and forces our gaze to the violence wrought upon nonhuman life.

The Dobbs Lane image is paired with an indistinguishable piece of artwork which Baker describes as a drapery from the medieval period. Baker's pairing of each image of roadkill with another image of either the surrounding landscape, church artwork or different fragments of the environment serves to anchor each photographed corpse to a place and time. The photos of each dead animal illuminate and register the stranger with whom we have never became acquainted. In his work *About Looking*, John Berger makes a clear distinction between the private and the public photograph. Baker's image is clearly categorised as the latter, and therefore, the image is, as Berger suggests, a 'memory of the unknowable and total stranger' (Berger 1991, 56). Baker attempts to trace a sense of familiarity amidst this unknowability through the compilation of images with captions, date and montage—a work of memorialisation, and an act of mourning the unmourned. This act of anchoring and exposure of animal bodily vulnerability draws us again to Pick's argument for a creaturely poetics. Pick argues that 'vulnerability, then, does much more than argue for animal rights or the reduction of suffering. It brings another world into view in which animal are not food. As the frame shifts and perception transforms, different moral arguments are possible' (Pick 2018, 415). The shift in Baker's work is the movement from perceiving 'roadkill' as a natural inevitability of petro-infrastructure to a death which is worth mourning and attending to.

Baker's artistic practice also follows Berger's ambitions for a form of photography which embarks on alternative practices outside capitalist consumer images, a photography practice which can imagine alternative futures. Thus, if Baker's work is one of imagining alternative futures, it is one where the death of the nonhuman is not considered simply as collateral damage. Berger further outlines,

> Photographs are relics of the past, traces of what has happened. If the living take that past upon themselves, if the past becomes an integral part of the

process of people making their own history, then all photographs would require a living context, they would continue to exist in time, instead of being arresting moment. (Berger 1991, 41)

Baker is therefore not simply tracing moments of the past but initiating and enacting a lived encounter with the dead. As observers, we are confronted with a frozen image that imprints in our memory, and we are no longer able to turn away or forget. The task of creating a political and social photography for Berger is not just to record the rest of the world but to act as a recorder for those involved in the events photographed (Berger 1991, 62). To be the recorder for those involved is to offer a mode of address to the beings depicted in the image. Baker illustrates this through the interruption of the bike in each frame as tracing his own encounter with the animal corpse. There is no clear subject or object in each montage; instead, it is a record of an address, an encounter between different elements. Baker highlights the power of memory and mourning as he recalls in an interview stating the words of Angela Singer: 'The animal, having no grave site, no bodily burial, becomes its own memorial' (Singer; Baker 2012). Baker's montage of images thus offer this memorialising function creating a photography project which bursts out different forms of affect and reaches the observer to unravel the naturalised conception of roadkill as collateral damage. In the words of Berger, 'a radical system has to be constructed around the photograph so that it may be seen in terms which are simultaneously personal, political, economic, dramatic, everyday and historic' (Berger 1991, 67). These profound combinations of opposites find themselves in Baker's exhibition, as he opens and defamiliarises the strange from the mundane to the extraordinary, exposing the observer to the unique and forgotten traces of animal remains. Baker's photographic records serve as a portal to the nonhuman worlds—the lives and the deaths—which rarely intrude into our everyday reality. This artistic experiment explores peripheral spaces and bodies exposing the polarising conditions and response to petro-modernity and hyper-automobility.

The question of why we respond in certain ways to different events, visual stimuli and media has ethical dimensions and consequences. As Butler notes, 'ethics is less a calculation than something that follows from being addressed and addressable in sustainable ways' (Butler 2016, 181). Influenced by Emmanuel Levinas, Butler frames ethics as a mode of address and response, an ethics is foregrounded and made possible through an interaction and engagement with an Other. Baker's project in 'Roadside' is to create a mode of address, one in which we are open to respond to the death of the nonhuman. The response to the suffering and vulnerability of another, as both Butler and Levinas conclude in their work, is not necessarily an ethical one but can be wrought with violence and rejection. As Pick further notes, 'neither Levinas nor Butler envisages the self-other encounter as naturally harmonious. On the contrary, the threat of violence hangs over the encounter with alterity' (Pick 2018, 416). The encounter with the animal's death is shaped by this dual movement, where both care and recognition can emerge or alternatively a process of deflection and further violence. This complex duality we find in the affective dimensions of disgust.

The montage image titled Woodbastick Road, shown further below, moves us not necessarily to mourn the loss of life but to a sensation of revulsion at the animal body atomised and splattered across the road. The blood is fresh, maggots spewing out of its body, the insides completely exposed and revealed. Each limb is discernible but crushed from the weight of a tyre that had smashed the animal into the ground, splattering the insides of the creature's body across the cemented tarmac. The facial features are blurred amongst the blood and rot. What is displayed here is a grotesque display of the severe consequences of automobility, a horrifying and revolting image which is rarely captured. Disgust is often a blockage to an ethical response, it can be amoral, but in this case staying with the feeling of disgust becomes a way of envisioning of how we are affected by and respond to nonhuman life. Such a process of reframing envisions what Joensuu describes as 'when the disgust encounter is reframed as a meeting of subjects' (Joensuu 2019, 113). Disgust is therefore primarily to do with contact and exposure, coming into proximity with another.

To borrow Ngai's term, these 'ugly feelings' (Ngai 2009, 345) reflect how one disassociates from the suffering and cruelty oil culture inflicts. Disgust can result in averting of one's gaze, but to stay with and not turn away has powerful implications for understanding visceral and bodily affects and the ways in which they are structured according to hegemonic norms. Disgust, here, thus acts as a gateway to understanding how affect is directed and manipulated; if we explore these sensations of repulsion, we do not aimlessly devour an image with our eyes but become alert to what is before us. Mourning is not the only political tool and affective marker to instigate a recognition of nonhuman life and death. Exploring experiences of disgust invites a new terrain of interaction, recognising why we might avert our gaze in revulsion or why we may instead linger in fascination. In Ahmed's words, 'disgust is clearly dependent on contact' (Ahmed 2004, 85); and it is through these contact zones in which a creaturely fellowship may reveal itself as at once 'ridiculous, ungainly, carnivalesque even – but solid and unquestioning' (Pick 2011, 188). In Baker's work, disgust open us to bodies exposed over time to vulnerability and brute force, to rot and decay, subjected to pain and death (Fig. 4.2).

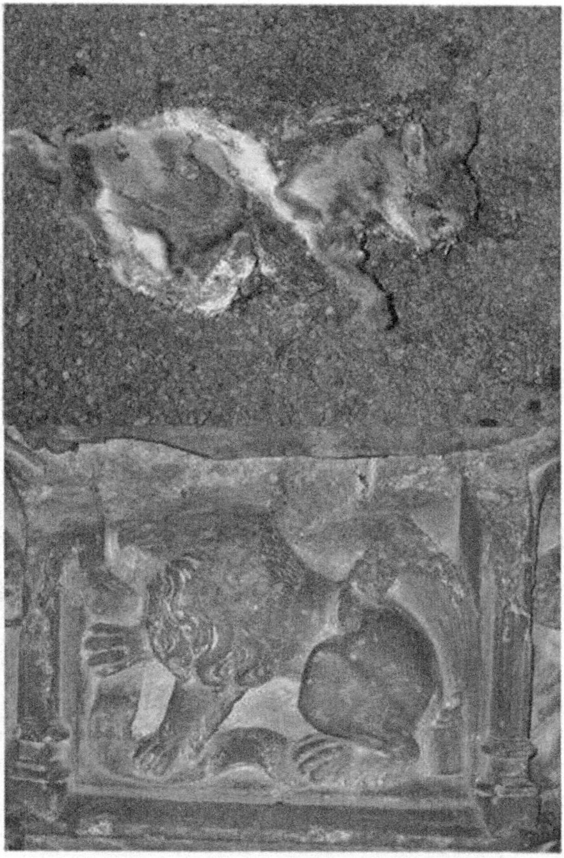

Fig. 4.2 Woodbastwick Road, 24/6/09

A further question generated by Baker's photography is the role of the observer: their political agency, to acknowledge, address and resist the violent mechanism of oil culture. As T. J Demos notes, in their exploration of visual culture and the Anthropocene, 'whatever we do, we cannot sit back passively and witness our own destruction as a source of visual pleasure or a neutral observer' (Demos 2017, 81). Does Baker's work induce an active engagement, or do we simply become a neutral observer? My exploration of disgust suggests otherwise to such a position of neutrality—the grotesque and poignant images are politically symbolic, registering and memorialising the deaths, the mutated and atomised bodies displayed in full rather than briefly acknowledged within our peripheral vision.

In his discussion of agony and photography, Berger claims,

> Confrontation with a photographed moment of agony can mask a far more extensive and urgent confrontation. Usually the wars which we are shown are being fought directly or indirectly in "our" name. What we are shown horrifies us. The next step should for us to confront our own lack of political freedom. In the political systems as they exist, we have no legal opportunity of effectively influencing the conduct of wars waged in our name. To realise this and to act accordingly is the only effective way of responding to what the photograph shows. (Berger 1991, 44)

Like Butler, Berger discusses photography in the context of war, analysing scenes of violence and suffering, exploring how these frames are manipulated and controlled to release certain desired affects. Baker captures a similar and everyday violence that is waged against nonhuman life but invites alternative affects that are not entirely bound or controlled. Berger's analysis in the above passage, however, points towards the limitations of the observer's response politically and socially to a photograph of violence and death. However, the act of address in Baker's work, the power of a gaze which is not shaped by power and oppression but of recognition, ignites a non-anthropocentric ethics, a way of mourning the lives which go unmourned. This in itself shapes a political movement to ethical action beyond the realm of the human, to not only mourn the loss and death of the nonhuman but register these creatures as a 'life' from the beginning (Fig. 4.3).

Baker's pairing of images of animal corpses with objects and landscapes of the surrounding environment invites the observer to situate and place

Fig. 4.3 Buxton Road, 17/11/09, Steve Baker

the lost life within a historical lived context. In the montage image titled Buxton Road, dated 17/11/09, the juxtaposition of frames includes an image of what could perhaps be a tail of the corpse of an animal. Once again, the taxonomy is unclear but what strikes the observer is the image it is paired with of the track marks of a tyre pressed into the ground. There is an attempt to recapture the event that led to this creature's death, searching for the traces of violence, illuminating the remnants of what could be coined a crime scene. Baker's artistic practice thus unravels the notion that 'roadkill' is a natural inevitability, the violence is left open and uncovered, and it is not buried or discarded but brought to the centre stage. Baker's photography forms, what Butler terms an 'intervention into the sphere of appearance' (Butler 2020, 202). The images convey 'the seriousness of looking' effectively intervening in the consensus around what is worthy of attention and recognition. In such a project of widening our sensory field, different trajectories of affect are formed, one which connects the observer and creates a mode of address to this nonhuman Other (Fig. 4.4).

The next image I wish to address is the ghostly apparition of what appears to be a dead bird, the wings spread and flattened across the tarmac, a strange spectre, haunting stains of what was a living creature. This image featured above, numbered VII on Baker's website, once more accentuates the encounter with Baker's bike, which is used as a framing device resting parallel to the corpse. On closer inspection, one recognises

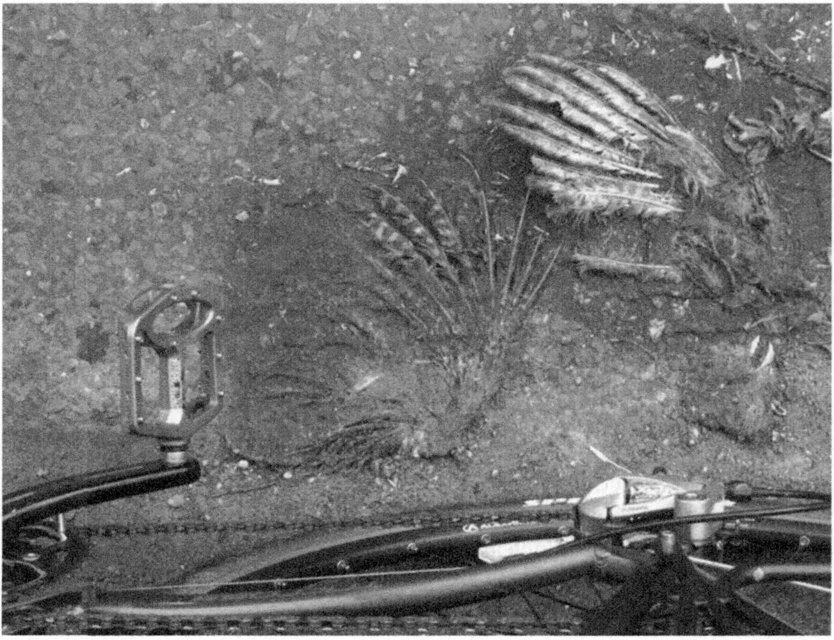

Fig. 4.4 VII, Steve Baker

the outline of the bird's face, its beak slightly protruding out from the ground—the wings, face and limbs are spread across the road. This atomised body is merged with the mud and soil, barely distinguishable from the ground itself. Without the employment of 'looking', of addressing and pausing one's attentive capacities to these animal remains this image would perhaps be lost; Baker renders visible a life and death on the precipice of disappearance—the stains of its body just about visible to both Baker and the observer. The act of address this image calls for resonates with Ahmed's recollection of bearing witness to her mother's pain. As Ahmed suggests, 'sometimes there would be nothing for me to do' (Ahmed 2004, 29), the only action she could take would be to bear witness, to recognise her mother's pain. Baker's photographs are also pleas to bear witness to the violence and suffering endured by the nonhuman, inviting the question of 'What about the pain of others? Or how am I affected by pain when I faced with another's pain?' (Ahmed 2004, 29). The contingency of pain: coming 'to feel that which I cannot know' (Ahmed 2004, 30), underlies Baker's images—a work in which the ordinary encounter with roadkill becomes an extraordinary plea to bear witness, to feel beyond the human.

The encounter with Baker's roadkill images is an encounter with vulnerability moving us to a spectrum of feelings from grief, disgust and a shared sense of pain. An exposure to animal vulnerability reveals the power relations inherent within energy infrastructure, as a space that is organised and orientated around the human. Baker reveals who is left and abandoned at the roadside, forcing us to encounter the leftovers and animal remains which we leave in our wake. Outside the sealed enclosure of the automobile, Baker's photography orientates the spectator towards the vulnerability of nonhuman life. By anchoring nonhuman deaths, placing them within our field of vision, Baker foregrounds a non-anthropocentric form of mourning and fellowship with other creatures. Baker's project is therefore one that can be aligned with the ambitions of a creaturely poetics, a work of attunement, regarding the expansion and limits of attention. The creaturely is thus concerned with the exposure and attention to vulnerability, as Pick notes 'most importantly, perhaps, vulnerability is the tug of reality, an attunement to "the difficulty of staying turned … toward flesh and blood" (Diamond 2003, 77)' (Pick 2018, 422). Baker's photographs tug at the unbearable reality of the violence wrought upon the nonhuman. Exploring Baker's collection of animal remains gathers the remnants and forgotten lives impacted by petroculture, opening us to the disastrous yet normalised aftermath and consequence of automobility. But as Matthew Calcaro reminds us 'to attend to roadkill is to be reminded of this basic (and shared) structure of earthly existence' (Calarco 2023, 24). My methodological premise of a vulnerable and earthly ethics is brought to the fore here when we attend to the image of nonhuman death in Baker's collection.

Oil Spills in the Niger Delta in Helon Habila's *Oil on Water*

Exploring the collateral damage, nonhuman death and landscapes of capitalist ruin, we now move far away from Norfolk country lanes to the edges and devastated villages of the Niger Delta, a landscape awash with black oil in Helon Habila's literary account of the neo-colonial presence of the oil industry in Port Harcourt, Nigeria. Combining the incidents of road kill on Norfolk lanes with that of oil spills from across the globe demonstrates the all-encompassing but varying consequences of petro-modernity. Like my analysis of Baker's photography, affect, disgust and experiences of grief beyond the human come to the fore in my exploration of *Oil on Water*. Returning to a central question in Butler's oeuvre of grief, in my analysis of Habila's work I ask 'Whose life appears as a life, and whose loss would register as a loss? How does that demographic imaginary function in ethics, in policy, and in politics?' (Butler 2020, 63). This statement in some ways is the topic of Habila's *Oil on Water*: the narrative follows two journalists, Rufus and Zaq, as they search for the abducted wife of a British oil engineer by rebel militants. This narrative journey of abduction, however, is not as central to my reading but it is instead the description of the oil-soaked landscape, ravaged by flares and continuous extraction, floating dead fish on the waters, and the silenced and dead fauna are my focus. Like Baker's rupture of what Butler terms the frames of recognisability, Habila too bursts the controlled frame around what is looked at and what is buried and forgotten by providing a reframing, turning our attention to the Niger Delta. As I move through Habila's descriptions of pipelines bursting through rivers and villages of the Niger Delta, I point towards the role of grief, disgust and pain as simultaneously creating a mode of address, furnishing a political community, and gesturing towards a creaturely ethics which encompasses and planetary and earthly ethics.

Habila's novel emerges from a rich literary tradition in Nigeria of environmentalism and activism. The figure of Ken Saro-Wiwa who was persecuted and killed for his environmental struggles for Ogoni people is representative of the coalition between literary form and environmental activism. In her work, Fiona Moolla considers how 'the intersection of the art and activism of Ogoniland's Ken Saro-Wiwa, executed in 1995 at the behest of dictator Sani Abacha, is generally considered in the context of postcolonial resistance to predatory oil exploitation' (Moolla 2016, 3). Ken Saro-Wiwa's death in the face of struggle against multi-national oil companies in the Nigerian region points to the sense of urgency the poetic and literary form hold in the fight against extraction. The Nigerian poet Nnmino Bassey's work, 'we thought it was oil, but it was blood' (Bassey 2002, 2), is emblematic of a violent struggle, demonstrating 'how poetry, indeed literature, can push the boundary towards activism' (Egya 2021, 5).

Nigerian literary history is immersed in a battle to salvage the Niger Delta landscape from further extraction. As Sule E. Egya notes, 'the Delta region, till today, remains the most sordid example of de-naturing of Nigerian landscape, where the effects are vivid and unashamedly left unclean – oil spills, gas flares, abandoned large equipment, polluted waters, etc.' (Egya 2021, 5). Habila's novel magnifies this landscape, drawing our attention to all different forms of life which are impacted by extraction regimes. Habila's narrative thus emerges from a literary tradition that is focused upon a drive for change, a sense of urgency and desperation to attend the crisis unfolding, attending to a history of colonisation and the neo-colonial landscape of extraction. The push towards literature to become a form of activism is evident in the novel's affective trajectories of grief, disgust and pain which force the reader to look and address oppression and environmental devastation.

In the opening passage of the novel, we learn that the central character's family has experienced 'an explosion in the barn with oil drums' (Habila 2011, 3). Although not explicitly related to the oil industry, Rufus comments: 'No, it was not a pipeline accident, as I told the white man, as I wrote in my published piece. But it might easily have been one, like in the countless other villages' (Habila 2011, 3). It appears, like the perceived mundanity of encountering roadkill accounted earlier, that oil spills and pipeline accidents are a normalised and habituated experience for the residents. Oil spills are thus seen as a form of collateral damage, a naturalised inevitability for a region in the grasp of the oil industry. But, like Baker, Habila challenges such notions and the frequent normalisation of violence through a magnification of crude oil's devastating effects. In both cases, Baker and Habila form what Imre Szeman describes as 'the impact of energy on literary [and visual] form at the periphery' (Szeman 2017, 281). Baker and Habila embark on a project of rendering visible what can be often viewed as the marginal/peripheral impacts of oil culture, or as Maximillian Feldner has described it, 'the dark underside of the oil business' (Feldner 2018, 518). The dark underside is thus also a project of coming to feel that which I cannot know that is articulated by Ahmed, a process of acting upon what we have not experienced but which needs to be addressed and understood.

Like Baker's images of grotesque, mutilated and rotting bodies struck by oncoming automobiles, there is also a sense of visceral corporeality to Habila's descriptions of the oil ravaged landscape. As Feldner has noted, Habila's 'haunting depiction of the Niger Delta's environmental destruction is heightened through rhetorical devices such as the personification of the landscape as a sick and dying person' (Feldner 2018, 516). Not only is the land personified but crude oil itself appears as a violent actor. In Michael Niblett's discussion on the literary representations of crude oil and sugar-cane he argues that,

> Oil, too, has known personification in literary representation. Indeed, its power to transform societies and energize global infrastructures means it has often appeared as more than an historical personage [...] oil is frequently presented as an all-powerful demon or god. (Niblett 2015, 273)

Habila's personification of both the land and crude oil illustrates the aesthetics of life at the periphery, exposing the horrors of commodity extractivism.

Nonhuman death and a dying ecosystem haunt Habila's text making nonhuman grief a prevalent feeling. The sense of spectrality we found in Baker's work; the haunting aspects of animal remains we also find in Habila's work. In the following description, Habila develops an affective sense of grief and despair:

> The next village was almost a replica of the last: the same empty squat dwellings, the same ripe and flagrant stench, the barrenness, the oil slick, and the same indefinable sadness in the air, as if a community of ghosts were suspended above the punctured zinc roofs, unwilling to depart, yet powerless to return. (Habila 2011, 9)

The striking image of a community of ghosts, suspended, unwilling to depart, yet powerless to return captures a community that is trapped physically yet on the precipice of disappearance; they are, in every sense, ghostly. Just like the body of the bird in Baker's image VII, haunting remains are present, yet it is a life or lives which are lost. Simultaneously, Habila combines images of ghostly apparitions with a rotting and diseased body to describe the landscape, 'the patch of grass growing by the water was suffocated by a film of oil, each blade covered with blotches like the liver spots on a smoker's hands' (Habila 2011, 9). The allusion here to a smokers hands covered with liver spots renders the earth a corporeal entity, gradually decaying and diseased through contamination and polluting substances. The frequent reference to the 'oilfecund earth' (Habila 2011, 34), 'the oil-scorched earth' (Habila 2011, 75), imagines oil as a suffocating and all-encompassing substance—where, in some ways, the commodity itself becomes representative of neo-colonial oppression that pervades the region. It also demonstrates a planetary aesthetic which is akin to an ethics of vulnerability where the earth is fragile and perishable, like our own bodies, vulnerable to force and pain.

Habila captures the histories and biographies of violence wrought on the Nigerian landscape not only through a devastated ecosystem and an almost gothic register but also in the complex figures of the rebel militants. The leading rebel militant, called the Professor, articulates the controversies over land ownership and how the Nigerian people have been dispossessed and marginalised. He urges Rufus the journalist to tell the oil industry figure heads: 'how we are hounded daily in our land. Where do they want us to go, tell me where? Tell them, we are going nowhere. This land belongs to us. That is the truth, remember that. You can go' (Habila 2011, 210). This is a provocative and urgent reminder of histories of the coalescence between land dispossession through colonialism and the ongoing neo-colonial forces of extractivist industries. As the Professor insists, 'tell them about the flares you see at night, and the oil on the water' (Habila 2011, 210). This imperative to tell them, to show them, whoever this 'them' might be, demonstrates how *Oil on Water* is, like Baker's project, a work in which rendering visible, widening one's sensory field, and magnifying the 'peripheral' impacts of oil culture is central to the narrative's motivations. Both Baker and Habila embark on this project of attention, exploring the expansions and limits of attention, asking the reader or spectator to bear witness to violence and the unbearable vulnerability bearing witness exposes.

In this project of rendering visible, Habila provides a dystopian, almost sci-fi, depiction of the Niger Delta—the invading alien life form, however, is in this case the colonising force of the oil industry. As Habila describes,

> It looked like a setting for a sci-fi movie: the meagre landscape was covered in pipelines flying in all directions, sprouting the evil-smelling, oil-fecund earth. The pipes criss-crossed and interconnected endlessly all over the eerie field. We walked inland, ducking or hopping over the giant pipes, our shoes and trousers turning black with oil. (Habila 2011, 34)

In such a description, sci-fi, and the notion of speculative fiction in particular, is envisioned as something which is not distant or far in the future, on another planet or existing within a different time zone; the apocalypse has instead already arrived. The pipes criss-crossing endlessly suggest a labyrinthine structure, oil culture thus appearing as an inescapable oppressor, the black oil entirely all-consuming, covering their clothes and bodies, which are entrenched within the soil and water. Oil is described as producing an inescapable stench, as if it is a rotting body, a corpse left to decay in the open. Habila permeates the entire sensory field with oil, 'the foul smell that clung to our hair, and itch on our grime-smeared faces' (Habila 2011, 10). To return to what Ngai describes as 'ugly feelings', we are reminded of the affective power of disgust. Like the visceral affects provoked from Baker's grotesque images of animal corpses, there is within Habila's literary imagery an evocative experience of disgust and repulsion. If, as many aesthetic critics have noted, disgust's origins lie within the role of eating and what is indigestible, the fear of contamination and disease, Habila's representation of the landscape of the Niger Delta is indeed rotten and decayed, the ecosystem devastated and unable to sustain its communities of human and nonhuman life.

For Amanda K. Greene, she suggests that 'disgust offers an alternative, affective way of encountering the pain of others that still challenges the more soothing logic of mourning and meaning-making' (Greene 2018, 483). Disgust evoked by the 'oil fecund earth' in *Oil on Water* is therefore symbolic of an earth which has become the abject yet suffering body, ravaged and looted, and left to rot. Habila renders visible the suffering of the Nigerian community through disgust. Greene further describes the affective potentiality of disgust: 'its invasive physicality collapses distance even as it instates it and undoes aesthetics even as it is artistically rendered. While the casual utterance of disgust may be a shield, its visceral representation and real affective force do not cover up traumatic events but instead stick to them and to the bodies of viewers that cannot remain disinterested' (Greene 2018, 483). Like Ngai, Greene demonstrates the critical productivity of disgust relating not just to the domain of politics but that of ethics too, of conveying another's pain and suffering. As Greene suggests, by disgusts eradication of distance one is forced to recognise trauma and to not deflect or turn away. The reference to a diseased and abject body recurs throughout the novel. When Rufus contemplates the fate of the white woman he questions,

> What could fate possibly want with her on these oil-polluted waters? The forsaken villages, the gas flares, the stumps of pipes from exhausted wells with their heads capped and left jutting out of the oil-scorched earth, and the ever-present pipelines, criss-crossing the landscape, sometimes like diseased veins on the back of an old shrivelled hand, and sometimes in squiggles like ominous writing on the wall. (Habila 2011, 175)

Imagining the pipelines as diseased veins on the back of a shrivelled hand once again captures a moment and response of disgust, the pipelines' labyrinthine structure depicted as suffocating and oppressive, thus rendering visible the pain of the earth. Disgust, for Greene, penetrates the body; it is a bodily rejection, a gag, a repulsion, and in this sense a bodily intrusion. The imagery of disgust perhaps conveys a wider signification of an invasion, not only of the body but of the earth and land of Nigeria, the intrusive violence of colonialism and the despotic monopoly of Western extraction regimes. Disgust conveys a physical and psychical pain, a process of decline and decay. It is, as Aurel Kolnai describes, a 'phenomenon of decomposition' (Kolnai et al. 2006, 486). Habila, however, allows one to stay with the experience of disgust. Such an experience, as Greene highlights, is 'to be forced to linger in disgust is to be trapped in proximity with, and physically invaded by, ghosts banished from the symbolic machine' (Greene 2018, 486). This sense of proximity, of being trapped among the ghosts, is to linger among the remains of oil culture, facing the consequences of violent extraction. It is clear that both Baker and Habila's project explores this politics of visibility, of attuning ourselves to an attentive capacity, a way of experiencing disgust without averting one's gaze. In Winfried Menninghaus' detailed exploration of disgust, he suggests 'the fundamental schema of disgust is the experience of a nearness that is not wanted. An intrusive presence, a smell or taste is spontaneously assessed as contamination and forcibly distanced' (Menninghaus 2003, 1). In Baker's and Habila's work, we are forced to reconcile and embrace this experience, attuning to what is usually expelled from both the body and from our field of vision. This ability to refuse to deflect recalls Pick's analysis of the creaturely, in which the encounter with vulnerability can lead to either exposure, a creaturely fellowship or a process of deflection. Becoming open to the communicative pressures of earthly vulnerability, however, leads to a process of attunement to nonhuman life. It is, as Pick suggests, 'a rawness of nerves', it has become a wound (Pick 2011, 9). Disgust, in this sense, is like a wound, revealing similar corporeal boundaries, yet moving our attention towards the pain and suffering of another.

By introducing forgotten and neglected terrains, the decaying landscape, the rotting animal corpses, the oil that floats on water, Habila challenges the distribution of 'grievability'. As Butler argues, 'in the same way that we talk about the unequal distribution of goods or resources, I believe that we can also speak about the radically unequal distribution of grief' (Butler 2020, 75). We have explored the ways in which disgust brings a different kind of suffering to light, a way of visualising a nonhuman alterity, the corporeal elements of the earth itself as a diseased and polluted body. But how and in what way does Habila also introduce a way of attending to the deaths of nonhuman life forms, furnishing a capacity for a non-anthropocentric mourning? For Butler, 'grievability governs the way in which living creatures are managed, and it proves to be an integral dimension of biopolitics and of ways of thinking about equality among the living' (Butler 2020, 56). As we now explore Habila's attention to the nonhuman lives and deaths or those who are cast out of the category of the human, it is important to think of grievability as a biopolitical regime and the ways in which Habila addresses this.

Habila's novel is attentive to the absence of nonhuman life in what was once a thriving ecosystem. As they board a boat, floating across 'the black expressionless water' (Habila 2011, 10), Rufus, the narrator, observes that 'there were no birds or fishes or other sea creatures – we were alone' (Habila 2011, 10). In this description of floating across the Niger Delta, we become aware of the nonhuman deaths which, although not physically apparent, are visible through a declaration of their absence. As they follow a bend in the river, however, they begin to see 'dead birds draped over tree branches, their outstretched wings black and slick with oil; dead fishes bobbed white-bellied between the tree roots' (Habila 2011, 9). Like Baker's visceral encounter with Norfolk roadkill, the characters of the novel are similarly confronted with the aftermath of oil culture, the devastating effects of extraction—scenes which are in many cases rarely brought to view. Butler's question 'whose life appears as a life, and whose loss would register as a loss?' (Butler 2020, 85) is also provoked by Habila's imagery of absence and death. The Niger Delta becomes an open morgue in Habila's writing where 'strange objects would float past us: a piece of cloth, a rolling log, a dead fowl, a bloated dog belly up with black birds perching on it, their expressionless eyes blinking rapidly, their sharp beaks savagely cutting into the soft decaying flesh' (Habila 2011, 34). The countless corpses floating across the black expressionless water, the oil described at one stage 'like a hangman's noose round the neck of whatever life form lay underneath' (Habila 2011, 215) captures the dark and gothic underside of oil culture. The frequent allusion to the earth as a corporeal entity accentuates that the earth too is subject to injury. Exposing the damage that occurs across the Niger Delta opens up a realm of affects, disgust and repulsion at the oil-soaked waters, despair and loss at the deaths of different life forms and amidst this ruin, a demand for justice. By invoking various affective registers from disgust to grief, Habila not only records and attends to ecological ruin and nonhuman death but calls for a mode of resistance. Butler's concluding remarks in *The Force of Non-Violence* gesture towards such a demand for hope through attuning to affects, as Butler declares,

> So, whether we are caught up in rage or love – rageful love, militant pacifism, aggressive nonviolence, radical persistence – let us hope that we live that bind in ways that let us live with the living, mindful of the dead, demonstrating persistence in the midst of grief and rage, the rocky and vexed trajectory of collective action in the shadow of fatality. (Butler 2020, 204)

By moving through the wreckage, collateral damage, of the extraction regimes across the Niger Delta, Habila provokes a series of affective responses which introduce a mourning that does not lead to despondency or despair but an anger for change, for recognition, a movement for justice for the lives that fall out of the frame of significance. Exploring the range of textual affects, of bodily intensities produced from literary and visual material, allows one to not reject or turn away but rather to become immersed in the feeling, to be in tune with the suffering of another whether that be human, animal or plant. Baker and Habila invite what Nixon terms 'sights of the unseen' (Nixon 2013, 17) but also provoke different affective trajectories that can change the structures around what life is considered as a life and whose loss would register as a loss.

Habila's novel also echoes Ahmed's notion of the 'contingency of pain', drawing us into the suffering of another place and time, allowing the reader to become connected to that which they cannot know. For Ahmed, 'our question becomes not so much what is pain, but what does pain do' (Ahmed 2004, 27). Likewise, Habila's novel conveys pain not so much to explain the experience but rather to ask what pain can do. Habila's narrator, Rufus, contemplates the fate of the wife of the oil engineer who has been abducted by rebel militants: he wonders 'maybe fate wanted to show her first hand the carcasses of the fish and crabs and water birds that floated on the deserted beaches of these tiny towns and villages and islands every morning, killed by the oil her husband was helping to produce' (Habila 2011, 175). The passage contemplates the suffering and death of nonhuman life forms, the wreckage of the exploits of oil engineers and Western corporations. There is an emphasis on the term 'show' in the passage, of looking towards something which is often edited out of the frame. By becoming a witness to this pain, however, Habila provokes questions around what the witnessing of such suffering can do. To return to Ahmed's reflection on her mother's illness, she notes 'through witnessing, I would give her pain a life outside the fragile borders and much loved body' (Ahmed 2004, 30). Although the circumstances and context are drastically different, Habila too asks one to bear witness to the devastation and pain occurring within the Niger Delta so the events begin to have a life beyond its borders, a movement towards recognition and resistance.

Grief, disgust and pain provided an entryway into feeling beyond the realm of human in Habila's text, opening up our attentive capacities to events and suffering that is often framed as peripheral. This project of attention displaces 'man as the centre of the universe' (Pick 2011, 183) and instead places emphasis on what Pick terms as the creature, a temporal, corporeal and finite being invoking a planetary and earthly vulnerability. Habila illuminates the corporeality of earth itself subject to pain and violence and therefore exposing the nonhuman suffering which aims to deliver us beyond anthropocentrism to a shared sense of finitude and vulnerability with other creatures and nonhuman life forms.

Contaminated Waters and Indigenous Resistance in *Perpetual*

Discovering hope in scenes of ruin, awakening one's attention to the lives discarded in the face of extraction and an ever-expanding infrastructure reveals how generating different modes of address can be a radical act inspiring an earthly sense of kinship and relation. Literature and visual culture garner our critical attention, widening our senses, furnishing a capacity to see what another sees, to feel someone else's experience of suffering. The shared sense of finitude and vulnerability which a creaturely poetics encourages serves to widen the frame on how we perceive human and nonhuman life. The reconfiguration of the human as a creature among many, material and temporal, creates an expansive frame in which we attend, care and mourn the lives who are often edited out of moral and ethical frameworks.

Rita Wong's and Cindy Mochizuki's graphic novel *Perpetual* is attentive to the ways extraction regimes impact a variety of communities, how we are all entangled in oil culture in varying ways with differing consequences. Like Baker and Habila, Wong and Mochizuki are also engaged in a project of reframing and expansion, a work focused on turning our attention to the remains and debris of energy infrastructure. The graphic novel is thus engaged in creaturely expression, an artistic endeavour that asks about the limits of attention.

Exploring the contaminated and polluted waters of Indigenous communities in Canada and the damage of extraction in the tar sands, Wong and Mochizuki reveal both the severe consequences of petro-modernity and suggest modes of resistance to it. Revealing the power of water as a source that connects us all, Wong poetically articulates the need to register and to feel beyond our sensory field: 'keep breathing deeply – this helps us to be mindful and present, and attentive to how we are related to creatures and places we may not see' (Wong 2015, 15). This practice of attunement and attentiveness facilitates a creaturely fellowship and the vulnerable ethics I seek to engender.

As I now explore the debris and remains of oil culture across the pacific, the Athabasca River and the underwater streams of Vancouver, Wong's insistence on being attentive to what we may not see is an important aspect of my analysis, a way of understanding an interdependency that is not always evident. To recall Ahmed's ethical demand that 'I must act about that which I cannot know' (Ahmed 2004, 31), *Perpetual* similarly calls for us to come to feel and act upon that which does not belong to us.

First, before I turn to my reading of *Perpetual*, I want to focus on Rita Wong's activist work and reflections on the healing walks she carried out in the tar sands led by an Indigenous community. In doing so, I draw on Wong's own affective trajectory in relation to the land and how her experience leads to the creative work of resistance. Like Habila, Wong notes how colonialism is not a thing of the past but rather its legacies continue in 'virulent, violent forms, in both the slow poisoning of water and the fast killing of women' (Wong 2015, 103). This becomes all the more pertinent in the healing walks across the tar sands as she describes,

> Each time, seeing the devastation to the land triggers a sick feeling in my stomach. On my own, I think I would have shrunk down into despair or numbed myself because I felt incapable of addressing the huge, overwhelming scale of the destruction. Yet, on this walk, the sick feeling coexists with a quietly hopeful one, invoked by the efforts of my co-walkers, as well as the many people we know who cannot make the long journey to Fort McMurray, the epicentre of tar sands extraction in northern Alberta, but who ask us to carry their wishes and prayers for the healing of the land with us. (Wong 2013, 133)

This triggering experience, of seeing the land in a state of destruction, is met with a series of ambivalent feelings—the overwhelming despair at being powerless to change what has already been done, alongside an experience of hope invoked by collaboration and a community coming together to care for the land. Further on in her commentary, we find a resemblance to Habila's literary rendition of the oil-soaked landscapes of the Niger Delta, as crude oil becomes an all-consuming and suffocating substance:

> If you had come, you would have smelled the sulfur and tar, the toxic pollution that lingers the next day on skin and hair. You would have heard the propane powered cannons that pop to scare off the ducks from landing in the enormous pools of poisonous water. It smelled like I was farting tar, inhaling it, ingesting it and eventually excreting it, while immersed in the gaseous fumes that hovered over the tar sands basin. (Wong 2013, 134)

Once again, we find the corporeal metaphors of a diseased body, a sense of disgust, of a repellent substance contaminating and the body of the earth and its inhabitants. Co-existent with this theme of decay and disease is the central need to heal, following Indigenous knowledges of protecting and thus healing the land. Within this experience of the healing walk, Wong suggests that 'the will to heal is humble and tenacious, inviting creative responses, ideas and more actions' (Wong 2015, 137). Like *Oil on Water* and *Roadside*, *Perpetual* is an affective call for justice, a record of environmental devastation, and impels creativity as a mode of resistance and a passage to healing.

Wong's poetry begins by asserting 'water as poetics & praxis' (Wong 2015, 5), a way of imagining an interconnectedness through the lens of water, 'part of [the] hydrological cycle' (Wong, 5), 'this planet's crucial circulatory system' (Wong 2015, 5). As Wong goes onto describe, 'your brain is roughly 85 % water, your body approximately 70%' (Wong 2015, 5). In this opening passage, the human body is not isolated or atomised but made up of different elements, part of a hydrological system, embedded within passages and storages of water. Like Butler's scholarship on precarity, Wong illuminates an interconnectedness but through the lens of water. By doing so, she also provides a 'critique of anthropocentric individualism' (Butler 2020, 73) and as Hannah Boast has noted presents 'non-anthropocentric modes of relating to water' (Boast 2021, 1). In this vein, I'd like to push Wong's poetics further to suggest how it can possibly furnish both a creaturely mourning, a planetary vulnerable ethics and thus a new relation to energy beyond extractivist and colonial logic. Mourning becomes a political tool of recognition in the graphic novel, a mourning which attends to ruin and death yet aims to strive and care for the living. As in Wong's healing walks across the tar sands, the graphic novel allows us to become alert to alternative energy processes of water, sun and wind that are living amongst ruin. Wong describes the healing walks noting 'the wind often blows fierce and constant, and the sun peeks through the clouds and smog now and then, reminding us that other forms of energy exist right here, in Alberta, that are cleaner and equally powerful to the dirty tar' (Wong 2013, 138). There is a glimmer of hope and a dream of transition in Wong's and Mochizuki's creative work in *Perpetual*, grounding us in the embedded hydrological cycle that flows through our bodies, across polluted shores, rivers and oceans.

Perpetual takes us from the underground streams of Vancouver, streams that are paved over by cement, as she listens for 'the dips and cracks in the ground that might signal stubborn streams underneath' (Wong 2015, 8), and then moves to the great body of water that is the pacific ocean. In the illustration, Mochizuki captures the remains and debris, petro-plastics which float across the shore and the ocean bed. Wong describes the image as 'a human crime scene' (Wong 2015, 33), 'with marine corpses [...] bloated [...] by plastic pollution' (Wong 2015, 33). As with Baker's exhibition *Roadside*, Mochizuki and Wong record the countless nonhuman deaths that are a consequence of a world reliant on petroculture. In the following pages, we enter the depths of the ocean, words are scattered across the panels, capturing the debris which fills the ocean itself. 'Look down' (Wong 2015, 33) and we find an array of leftovers: 'roadkill [...] burger wrapper [...] ragged newspaper [...] oil smeared plate [...] a child's dropped shoe [...] straw [...] tinfoil [...] sock [...] crinkly cellophane' (Wong 2015, 37). Scattered across the page, these words are placed next to the panel of the ocean, where fish swim in and out between these leftovers and broken remains. Wong and Mochizuki illustrate and collect the remains of a consumer culture that is embedded in petrochemicals, 'the juice carton', 'the shopping bag'—all created from crude oil and scattered across the ocean bed. Here, we find a sense of grief for the environment in a state of ruin while invoking our own accountability and complicity in its devastation. This sense of complicity is further implicated when Wong articulates that she is 'living in the belly of the bitumen beast driving and flying, I am implicated in the oil addiction I critique' (Wong 2015, 57). There is no way to escape crude oil—a society driven by and shaped by this commodity. However, addressing its effects, and accounting for complicity, can therefore expose paths to transition.

This experience of both grief and complicity returns us to Butler's argument in *Precarious Life*. Butler proposes 'to consider a dimension of political life that has to do with our exposure to violence and our complicity in it, with our vulnerability to loss and the task of mourning that follows, and with finding a basis of a community within these conditions' (Butler 2006, 19). It is in such conditions, overwhelmed by grief, accountable and yet exposed to violence, in which Wong and Mochizuki situate environmental damage in their work. Wong asks a haunting and provocative question of 'how do you respond to [...] devastation?' (Wong 2015, 47). Behind these words, Mochizuki has illustrated a bird covered by black oil. The question is similar to that of Butler, how do we fashion a community in the face of violence and mourning? What are our modes of resistance to environmental devastation? Mourning for the land and environmental devastation here provoke questions, response and action. As the narrative turns to collectives and direct action among Indigenous communities, one activist declares that 'our current environmental reality is so urgent that we need to build alliances and relationships to work for the changes that are needed' (Wong 2015, 81). It is in this space of despair, vulnerability and mourning in which hope of collective action is built. This sense of urgency and exposure of vulnerability to different life forms further points towards Pick's creaturely ethics.

For Pick,

> Creaturely vulnerability opens up zones of "indistinction" (Calarco, 2015), where species identities blur and where different beings, or creatures, are perceived as corporeal and vulnerable. More radically still, creaturely vulnerability, as I understand it, calls for the contraction of humanity rather than its benevolent extension to nonhumans. Creatureliness—the state of being exposed to natural necessity and the ravages of power—does not call for the alleviation of vulnerability via gestures of "humanization" but for more profound forms of "dehumanization." The creaturely, then, is focused on unseating the structures of human exceptionalism (less on the generation of empathy). By imbuing materialism with a sense of reverence for everything that is, creatureliness encompasses all life, from animals to plants. (Pick 2018, 414)

Pick notes how a shared vulnerability from animals to plants is a way of building alliances, opening up zones of indistinction. Vulnerability, in this sense, becomes not a plea for paternalism but a path to resistance. As Juiletta Singh suggests in her work on dehumanism, 'we become vulnerable to other possibilities for living, for being together in common, for feeling injustice and refusing it without the need to engage it through forms of conquest' (Singh 2018, 21). Vulnerability here becomes a way of becoming in common, creating unlikely bonds and alliances. But how and in what way does *Perpetual*'s exposure of vulnerability and modes of resistance gesture towards a different relation to energy, to the world around us, beyond violent extraction regimes?

Sheena Wilson has argued that we are currently at a stage of an energy impasse that defines contemporary climate discourse in North America—and beyond. This stage is 'a result of an atrophy of imagination that blockades transformative action' (Wilson 2018, 377). For Wilson, this impasse and blockade provokes a series of questions 'what does energy do? What is energy for? What from the age of oil is not working? And, most critically, for whom is it not working?' (Wilson 2018, 377). Wilson goes onto suggest that in order to create a just and feminist transition from the age of oil there has to be introduction of what she terms Other knowledge systems including but not limited to feminist and Indigenous perspectives. Invoking the planetary and earthly vulnerable ethics I have sketched so far, we can perhaps invite a possible passage way through the blockade of transition.

Perpetual, as well as Habila's and Baker's work, addresses what Sharae Deckard terms 'the infrastructure of violence' (Deckard 2021, 7), where pipelines lead to dispossession, draining, as well as sites of potential obstruction and occupation (Deckard 2021, 7). *Perpetual* presents a site of occupation and resistance through the power of water, as both a material force and metaphor of interconnected collaboration. Water, in a metaphorical sense, breaks the blockade of imagination as the writers, illustrators and activists come together in the creation of *Perpetual*, 'untapping watershed mind' (Wong 2015, 67). In their creative process of drawing, writing and talking, Wong and Mochizuki describe it as 'nourishing work of a power shift, paradigm shift' (Wong 2015, 73). Acutely aware of the need to transition to different ways of being and living in the world, Wong declares that 'we live in a watershed moment' (Wong 2015, 74). Like Wilson, Wong and Mochizuki are aware of the urgency to move away from the age of oil and to resist the violent colonial logics of extraction. Through the variety of registers within their poetry and illustrations, we encounter this tumult of emotion, a rebellion on edge of revolt, both an address to lives lost and a call for justice for the living.

For Butler, mourning 'has to do with agreeing to undergo a transformation'—a transformation that cannot necessarily be charted or planned (Butler 2006, 21). Undergoing this transformative experience through ecological grief and despair, *Perpetual* asks what can arise within the condition of mourning. As Wong goes onto ask, 'what might a watershed moment look like?' (Wong 2015, 78). After journeying through ruin, becoming alert to the impending climate crisis, transformative action is built in the creation of the graphic novel. For Wong and Mochizuki, the crisis in Canada and across the globe demands 'us to be conscious, to be mindful of our actions, our thoughts – how we live our daily lives. It would require us to take up the challenge to respond to the crisis posed by mega projects like the tar sands' (Wong 2015, 78). Meeting with an Indigenous activist Dorothy Christian, Wong and Mochizuki record the moment they attend an environmental conference. For the protection of the Indigenous lands, Gan-ya-ge-haga, also known as Montreal, Dorothy declares 'who is going to stand next to me when an army tank is coming at me' (Wong 2015, 82). Mochizuki's illustration captures a crowd with all their hands in the air agreeing to stand by Dorothy's side. Like Wilson's claim for the incorporation of feminist and Indigenous perspectives in the coming energy transition, Wong and Mochizuki foreground resistance and healing in the work of the Indigenous communities in Canada. The question of the movement away from the age of oil is 'how to decolonize and re-Indigenize' (Wong 2015, 89). A question which I would further add is how, in Julietta Singh's terms, to dehumanise? Moving away from human exceptionalism and its narrative of conquest brings us closer to the domain of the creaturely, the shared sense of vulnerability which marks all life, plant and animal, and can therefore allow us to embark on new configurations of interacting and understanding our wider ecology.

Energy regimes are shaped and enforced by a colonial logic, an erasure and dispossession of land, as well as erasure of ways of being and living in the world. To embark on a trajectory away from violent extractivism, Wong and Mochizuki invite us to 'listen to the lands, the waters, the first peoples, whose everyday practices continue, despite colonial violence and the theft of Indigenous children from their families' (Wong 2015, 89). To return to Wong's quote at the beginning of my exploration of *Perpetual*: 'keep breathing deeply – this helps us be mindful and present, and attentive to how we are related to creatures and places we may not see' (Wong 2015, 15), we find that by journeying across the Indigenous lands of Canada we have learned of places and lives that have been erased, while illuminating the remains and debris that we ignore. *Perpetual* exposes the infrastructure of violence, the pipelines cutting across rivers, the plastics filling the Pacific Ocean, but also moves from a state of mourning to a position of action and resistance, generating hope within ruin. *Perpetual* is therefore a work of creaturely and earthly expression, foregrounding relations beyond species divisions and towards our environment.

Mobilising Affect: Paths to a Creaturely and an Earthly Practice

Baker, Habila, Wong and Mochizuki presented literature and visual media which provided a reframing of oil culture and infrastructure placing the nonhuman at the centre. Their focus on the periphery is a project of attuning our attention to the devastation which occurs beyond physical sight and our sensory field, rendering visible projects of extraction and its multivalent effects. These textual and visual sources thus display the effects of energy production and its infrastructure. As Michael Niblett suggests, 'where a commodity so overdetermines all facets of society, its influence on aesthetic practice will be correspondingly marked' (Niblett 2020, 271). Baker demonstrates the consequences of a region dominated by the automobile creating art which responds to its impact. Habila conveys the grief, despondency and disgust of land dominated by the oil industry in the Niger Delta. *Perpetual* likewise conveys the violence of pipeline infrastructure and offers creative acts of resistance.

They are all exemplary works of exposure igniting affects which allow one to feel beyond the sphere of the human. These affective encounters with oil spills and animal remains point towards different ways of living and engaging with the world beyond the violence of extraction. To create a creaturely relation to energy and embark on the vulnerable ethical practice I have proposed, in hopes of moving away from the age of oil, is a project of attending to violence that is often remote and obscured from view. For a world after and beyond oil, we thus must attend to the present violence of energy infrastructure and leave behind the fossil fuel economies extractive logic.

References

Ahmed, Sara. 2004. *The Cultural Politics of Emotion*. New York: Routledge.

Baker, Steve. 2012. Stains, Drains and Automobiles: A Conversation with Steve Baker about Norfolk Roadkill, Mainly. *Art & Research, A Journal of Ideas, Contexts, and Methods*.

Baker, Timothy C. 2019. *Writing Animals: Language, Suffering, and Animality in Twenty-First-Century Fiction*. Palgrave Studies in Animals and Literature Ser. Cham: Palgrave Macmillan US.

Bassey, Nnimmo. 2002. *We Thought It Was Oil—but It Was Blood: Poems*. Ibadan, Nigeria: Kraft Books.

Berger, John. 1991. *About Looking*. 1st Vintage International ed. New York: Vintage International.

Boast, Hannah. 2021. Borrowed Waters: Water Crisis and Water Justice in Rita Wong's *Undercurrent. Textual Practice* 35 (5): 747–767. https://doi.org/10.1080/0950236X.2020.1731587.

Brown, Elspeth H., and Thy Phu, eds. 2014. *Feeling Photography*. Online-Ausg. Durham: Duke University Press.

Butler, Judith. 2006. *Precarious Life: The Powers of Mourning and Violence*. Verso.

Butler, Judith. 2016. *Frames of War: When Is Life Grievable? (Radical Thinkers)*. Verso.

Butler, Judith. 2020. *The Force of Nonviolence: An Ethico-Political Bind*. London; New York: Verso.

Calarco, Matthew. 2015. *Thinking Through Animals: Identity, Difference, Indistinction*. Stanford University Press.

Calarco, Matthew. 2023. *Reflections on Roadkill Between Mobility Studies and Animal Studies: Altermobilities*. Palgrave Pivot. Cham: Palgrave Macmillan.

Deckard, Sharae. 2021. 'Waiting for the Master's Dams to Crack': Hydro-Dependency, Water Autonomy and World-Literature. *New Formations* 103 (103): 134–155. https://doi.org/10.3898/NewF:103.08.2021.

Demos, T. J. 2017. *Against the Anthropocene: Visual Culture and Environment Today*. Berlin: Sternberg press.

Desmond, Jane. 2016. *Displaying Death and Animating Life: Human-Animal Relations in Art, Science, and Everyday Life*. Animal Lives. Chicago; London: The University of Chicago Press.

Diamond, Cora. 2003. The Difficulty of Reality and the Difficulty of Philosophy. *Partial Answers: Journal of Literature and the History of Ideas* 1 (2): 1–26. https://doi.org/10.1353/pan.0.0090.

Egya, Sule E. 2021. *Nature, Environment, and Activism in Nigerian Literature*. Abingdon, Oxon: Routledge.

Feldner, Maximilian. 2018. Representing the Neocolonial Destruction of the Niger Delta: Helon Habila's *Oil on Water* (2011). *Journal of Postcolonial Writing*, April, 1–13. https://doi.org/10.1080/17449855.2018.1451358.

Freud, Sigmund, and Shaun Whiteside. 2005. *On Murder, Mourning and Melancholia*. Modern Classics. London: Penguin.

Greene, Amanda K. 2018. Dismembering Remembering. *Twentieth-Century Literature* 64 (4): 483–503. https://doi.org/10.1215/0041462X-7298996.

Habila, Helon. 2011. *Oil on Water*. London: Penguin Books.

Joensuu, Eleonora. 2019. *A Politics of Disgust: Selfhood, World-Making, and Ethics*. Routledge.

Kolnai, Aurel, Barry Smith, and Carolyn Korsmeyer. 2006. *On Disgust*. Chicago, IL: Open court.

Menninghaus, Winifred. 2003. *Disgust: The Theory and History of a Strong Sensation*. Suny Series, Intersections: Philosophy and Critical Theory. State University of New York Press.

Moolla, Fatima Fiona. 2016. *Natures of Africa: Ecocriticism and Animal Studies in Contemporary Cultural Forms*. Johannesburg: Wits University Press.

Ngai, Sianne. 2009. *Ugly Feelings*. Cambridge: Harvard University Press.

Niblett, Michael. 2015. Oil on Sugar: Commodity Frontiers and Peripheral Aesthetics. In *Global Ecologies and the Environmental Humanities: Postcolonial Approaches*. Routledge.

Niblett, Michael. 2020. *World Literature and Ecology: The Aesthetics of Commodity Frontiers, 1890–1950*. New Comparisons in World Literature. Cham: Palgrave Macmillan.

Nixon, Rob. 2013. *Slow Violence and the Environmentalism of the Poor*. First Harvard University Press paperback edition. Cambridge, MA; London: Harvard University Press.

Out of the Woods Collective. 2020. *Hope against Hope: Writings on Ecological Crisis*. Brooklyn, NY: Common Notions.

Pick, Anat. 2011. *Creaturely Poetics: Animality and Vulnerability in Literature and Film*. New York: Columbia University Press.

Pick, Anat. 2018. Vulnerability. In *Critical Terms for Animal Studies*. Chicago University Press.

Sands, Danielle. 2022. *Animal Writing: Storytelling, Selfhood and the Limits of Empathy*. Crosscurrents CROSS. Edinburgh: Edinburgh University Press. https://doi.org/10.1515/9781474439053.

Singh, Julietta. 2018. *Unthinking Mastery: Dehumanism and Decolonial Entanglements*. Durham; London: Duke University Press.

Sontag, Susan. 2001. *On Photography*. Picador.

Stanescu, James. 2012. Species Trouble: Judith Butler, Mourning, and the Precarious Lives of Animals. *Hypatia* 27 (3): 567–582. https://doi.org/10.1111/j.1527-2001.2012.01280.x.

Szeman, Imre. 2017. Conjectures on World Energy Literature: Or, What Is Petroculture? *Journal of Postcolonial Writing* 53 (3): 277–288. https://doi.org/10.1080/17449855.2017.1337672.

Wilson, Sheena. 2018. Energy Imaginaries: Feminist and Decolonial Futures. In *Materialism and the Critique of Energy*. MCM Publishing.

Wong, Rita. 2013. Ethical Waters: Reflections on the Healing Walk in the Tar Sands. *Feminist Review* 103 (1): 133–139. https://doi.org/10.1057/fr.2012.21.

Wong, Rita. 2015. *Perpetual*. La Vergne: Harbour Publishing Co. Ltd.

CHAPTER 5

Transitions: Energy Futures and Multispecies Relations

The After Oil Research Collective argue that 'a genuine global transition away from fossil fuels will require not only a reworking of our energy infrastructures, but a transformation of petroculture itself' (After Oil 2016, 9). In this statement, the collective refer to how petroculture has come to shape values, feelings and societal norms; their argument is therefore that a cultural and social transition must take place to conceive of a future free from violent and imperialist modes of resource extraction. In Ursula Le Guin's essay 'World-Making', she contemplates world-making, a practice that ranges from envisioning and creating the world anew from religious imaginaries to the planets of science fiction and mere survival. If SF is an act of world-making, not just a focus on the new but about living with the old, how might energy and energy transition be reconceived within the SF imaginary? For Le Guin, science fiction is not so much about the future or capitalist imperial notions of progress but instead about our perception and awareness. As she suggests, 'as a science fiction writer I personally prefer to stand still for long periods, like the Quechua, and look at what is, in fact, in front of me: the earth; my fellow beings on it; and the stars' (Le Guin 2018, 140). I contend not only with envisioning future ways of being with energies and other species through the lens of SF and storytelling but to figure ways of remaking and living with the old—exposing the relations that we have which may have been broken or forgotten. What

J. Taylor, *The Nonhuman Narratives of Energy*, Palgrave Studies in Animals and Literature,
https://doi.org/10.1007/978-3-031-94332-4_5

relations become possible if mastery and progress are not their defining feature? Can a world after oil be imagined without the extractive regimes of capital marching forward? We witness in After Oil's more recent work on solarities, how they contend with the importance of storytelling and world-making for a just move away from fossil fuels. As they describe:

> We require stories that move away from solitary, individual heroes to multispecies stories that are grown over time, stories that are intertwined with other beings and celebrate not individual feats but the mutual creation of new ecosystems. To thrive collectively requires listening, learning, and making collectively. (After Oil 2022, 66)

The cultural works I explore challenge, enforce and illuminate different ways of storytelling that are in some cases solitary endeavours of transition and in others a collective struggle for attention and change to our environment and our nonhuman companions.

Graeme Macdonald in his work on energy and SF argues how a world 'After Oil' in fiction can restage the violence of the present through space operas, post-apocalyptic visions and resource depletion. The problem of the future then is perhaps dealing with and surviving in the present. The literary texts focus not only on envisioning the future but also on past and current relations and how we perform them in more just and ethical ways. The first text I turn to in this chapter is the short story 'Moonlight' by Cixin Liu from the Chinese SF collection *Broken Stars*. Here, energy transition is enforced by techno-fixes and green capitalist solutions. In this post-apocalyptic vision of the future, the violence which shaped the age of oil continues despite the introduction of renewable resources. Following this reading, I focus on Ursula Le Guin's *The Dispossessed* whose future worlds illuminate multispecies solidarity and present an alternative to the capitalist failures of 'Moonlight'. In my analysis of Le Guin's text, Weil's philosophy on affliction informs my reading of the scenes of suffering and solidarity in *The Dispossessed*. By thinking ecologically with Le Guin and Weil, I suggest that the novel has implications for energy transition. The final novel I explore is Stanislaw Lem's *Solaris*, which introduces life worlds beyond human mastery and resource violence; Lem imagines an interaction with and engagement across species lines towards interstellar regions and alien life forms. In my reading of these SF narratives, I combine animal studies scholarship with energy humanities, addressing the ways the two can inform one another and inspire a transition from the violence of capitalist production and extraction.

Throughout my exploration of the latter two SF narratives, I elucidate the ethics of an earthly vulnerability but with a particular focus on Julietta Singh's *Unthinking Mastery*. Alongside what Singh terms 'nonmasterful politics' (Singh 2018, 15), I draw on Anat Pick's notion of 'letting be' and core concepts from the philosophy of Simone Weil including decreation, attention and affliction. These theoretical and philosophical frameworks look to vulnerability as a way to establish an ethical position, a way of recognising the suffering of another. My project is therefore to suggest how all three theorists not only introduce a multispecies ethic through modality of vulnerability but also how they can inform just, feminist and perhaps even a 'creaturely' transition from the age of oil. If we are to move from violent modes of extraction, based on human mastery and capitalist logic, the 'human' itself must be brought into question. As Singh suggests, 'we can begin to address how drives towards mastery inform and underlie the major crises of our times – acts of intrahuman violence across the globe, the radical disparities in resources and rights between the Global North and the Global South [...] escalating threats of ecological disaster' (Singh 2018, 3). What Singh, Pick and Weil offer is a way of rethinking multispecies relations and our position to consider both our own finitude and temporal condition, discovering a sense of vulnerable co-existence and cohabitation with other elements and different species. My suggestion is that this perspective informed by these three thinkers can enable a different form of knowledge outside capitalist logic to embark on a transition away from contemporary violent energy regimes.

Science fiction is immersed in energy politics, legacies of colonialism, future dystopic visions of resource depletion and ecological collapse. I look to how science fiction offers what Gerry Canavan describes as a 'polyvocal archive of the possible' (Canavan and Robinson 2014, 16), presenting alternatives to the current imperial and capitalist system. I argue that science fiction is a world-building project which allows us to challenge the violence and inequalities of the present while articulating and envisioning different futures from the age of oil.

Science Fiction and Resource Politics

The questions raised in the research project After Oil serve as significant pointers to the ways in which science fiction relates to energy and resource politics. The first is 'how is the use of energy entwined with representations and narratives about modernity and the environment? Correspondingly, how do artistic productions reflect, critique, and inform our understanding and use of energy?' (After Oil 2016, 10). And lastly, 'what range of scenarios is currently on the table for imagining our future with energy?' (After Oil 2016, 10). Science fiction is particularly attuned to resource politics, colonial rule and energy regimes, not only in sketching the horrors of capitalist exploitation but possible alternatives to it. In my exploration here of energy culture and science fiction, I focus on science fiction's relation to colonialism, human animal relations and environmental injustice and examine how they are all connected to capitalist resource politics as they are explored in the SF imaginary.

The propelling force of capitalist progress and the search for more commodity frontiers can often be mirrored in the SF expedition to discover new worlds—space as the final frontier for mining and extraction, to conquer and invade. Gerry Canavan describes late capitalism as,

> a mode of production that insists (culturally) and depends (structurally) on limitless expansion and permanent growth without end: into the former colonial periphery, into the peasant countryside, through oil derricks into the deepest crevices of the earth, and, then, in futurological imaginings, to orbital space stations, lunar cities, Martian settlements, asteroid belt mining colonies, sleep ships to Alpha Centauri, and on and on. It is a process of growth whose end we can simply not conceive. (Canavan and Robinson 2014, 5)

Capital's path in SF can be a one-way journey of progress, without limitations, expanding further across space, to intergalactic regions. Capitalist imperialism's drive for limitless expansion is reflected in some SF tropes, presenting an arena of power relations, technological advances, and the appropriation of new land and resources. In John Rieder's important work *Colonialism and the Emergence of Science Fiction*, he explores this connection of the ideological drives of imperialism and its influence on the aesthetics and themes of SF. He begins this work with a quote from Edward Said: 'the novel as a cultural artefact of bourgeois society, and imperialism are unthinkable without each other' (Said 1994, 70–71). This relationship between the novel and imperialism is further developed in Rieder's work by suggesting that the emergence of science fiction arose at a particular moment of colonial history:

> Increasing popularity of journeys into outer space or under the ground in the late nineteenth and early twentieth centuries probably reflects the near exhaustion of the actual unexplored areas of the globe-the disappearance of the white spaces on the map, to invoke a famous anecdote of Conrad's. Having no place on Earth left for the radical exoticism of unexplored territory, the writers invent places elsewhere. (Rieder 2008, 4)

This connection to imperial power and SF's creation and invasion of new worlds is further elaborated by Michelle Reid who draws on Istvan Csicery-Ronay Jr., who goes further than Rieder, arguing that 'sf is itself a fantasy of empire, emerging in industrialised, imperial nations, such as America, Britain, France, Germany, Japan, and the Soviet Union, as a mediator between national cultural traditions and the rise of global capitalism' (Reid 2009, 258). The relationship between SF and the rise of global capitalism is something I wish to draw out further. Returning to the discussion of the 'human', colonial legacies and capitalism, there is clear illustration of these connections in science fiction as fantasies of invasion and appropriation of land are played out within the SF imaginary.

To move to more contemporary readings of science fiction, however, the eco-apocalypse genre often presents resource wars and energy depletion as its central narrative arc. Moving from the rise of capitalism with space operas and mass-market shows such as *Star Wars*, we now find ourselves immersed in an arena of ecological demise, dystopic visions and entropic worlds like the film *Mad Max Fury Road*, or Margaret Atwood's *MaddAddam* trilogy, post-oil futures focused on mere survival. Fictions such as Octavia Butler's *Parable of the Sower*, Paolo Bacigalupi's *Wind Up Girl*, Kevin Barry's *The City of Bohane* present worlds in which the famous quote of Fredric Jameson, 'it's easier to imagine the end of the world than it is the end of capitalism' (Jameson 1996, xii), is culturally materialised. These works show less of a utopian vision of energy systems and resource politics and more of their destruction and implosion. As Macdonald further notes:

> A characteristically ironic feature of eco-apocalypse narratives is their detectable strain of regret for the comforts and security of the petrolife we are now

> struggling to maintain: a "past" life, a post post-peak of continually desired or extended oil, most notably in its depletion, and especially after the event of an oil-driven collapse. (Macdonald 2016)

With this melancholic strand that mourns the comforts of capitalist and petrocultural society, the eco-apocalypse genre opens us to the chaos that possibly ensues in a world without oil. Is there, however, as Mark Fisher's *Capitalist Realism* argues, an alternative? As Canavan has asked, how might science fiction provide the 'polyvocal archive of the possible' (Canavan and Robinson 2014, 16) envisioning futures that are not tied to colonial and capitalist ventures or ecological demise but transitions to just and ethical energy futures?

SF can offer these modes of critique and alternatives by its engagement and encounter with the animal and the alien. To envision a world free from the violence of oil culture and capitalist logic in the literary and cultural imagination, I suggest that science fiction's focus on animal alterity is perhaps a key starting point. Sherryl Vint's *Animal Alterity: Science Fiction and the Question of the Animal* provides an effective insight into how SF explores the importance of nonhuman agency. As Vint notes, 'sf offers a wider scope than does most literature for enabling animal agency to become part of the quotidian world, as well as space to attempt to grasp animals as beings in their own right rather than as beings defined through their place in human cultural systems' (Vint 2010, 6). My contention is that science fiction's focus on nonhuman actors and different life forms can serve to create an ecological awareness, pointing to our vulnerable co-existence with other species. The violent disparities of resource distribution and depletion in the eco-apocalypse strain of SF is also often combined with an encounter between different life forms—both alien and animal, a life form that appears as an Other. What are the implications of this multi-species coalescence for energy futures? Can such powerful rethinking of the human in the interaction with different species call for a renewed energy ethic? If we are now at a current energy impasse, a deadlock, in which imaginations are blocked, how can science fiction itself contribute to ways of imagining transition? For the After Oil Research Collective, transition is not just a case of moving from one form of energy to another. As they declare, 'we will not make an adequate or democratic transition to a world after oil without first changing how we think, imagine, see, and hear' (After Oil 2016, 41). Changing how we think, imagine, see and hear has to do with relations and how we perform them. Literature, and science fiction in particular, pays attention to and critiques present power relations and can show alternatives to the contemporary violence of oil culture. Science fiction's engagement with an animal alterity has the potential to rethink human mastery and thus provides a template for our future interaction with our wider ecology from other species to energy sources. In what follows, however, I turn to Cixin Liu's short story 'Moonlight', addressing the failure of undergoing energy transition within the existent extractive mode of capitalist logic.

Green Capitalism and Violent Transitions in Cixin Liu's 'Moonlight'

Cixin Liu is a prominent Chinese SF writer, whose famous trilogy, *The Three Body Problem*, illustrates anxieties over extra-terrestrial invasion and presents the fears around nuclear energy. Liu is one of the contributors to Ken Liu's *Broken Stars* collection of Chinese Science Fiction. In Liu's contribution, the short story 'Moonlight' explores the rise of green capitalist solutions to climate catastrophe and the problems which arise from the fossil fuel economy. Liu himself comments on the limitations of technological advances on Earth to survive ecological disaster:

> There is no apparent short-term benefit to human terrestrial beings; on the contrary, even those endowed with greater governmental foresight are confronted with imagining the unique dilemmas presented when we imagine the successful establishment of extra-terrestrial colonies. (Liu 2019, 25)

For Liu, the solutions and technology needed to prevent climate crisis cannot be sourced on earth and instead he promotes intergalactic travel as a solution to planetary issues. 'Moonlight' stages these limitations of advanced technologies and renewable energy resources, where the outcome is bleak inaction, a blockade to energy transition resorting back to the fossil fuel economy. Green capitalism offers what is termed as 'sustainable' solutions to the ecological crisis. As Jacklyn Cock argues, 'The two pillars on which "green capitalism" rests are technological innovation and expanding markets while keeping the existing institutions of capitalism intact' (Cock 2011, 45). Liu presents the fatalities and failures of relying on capitalist solutions to the problem of energy and energy transition. David Schwartzman declared China as a world leader in renewable energies offering a vision of hope for the future. Liu's position in 'Moonlight' illustrates a less utopian view, where we discover the limitations of technological advances which are entrenched in global capitalism.

'Moonlight' explores the transition to renewable energy resources but envisions the dystopic and apocalyptic visions of a global system under the same imperial and violent modes of production that shaped the age of oil. In this short story, the protagonist, an energy planner, receives a call from the future, the year 2123, from his future self. His future self informs him that Shanghai has been flooded, 'the last of coastal cities to fall' (Liu 2019, 57), as climate change has ravaged the earth, 'the polar ice caps have gone' (Liu 2019, 57). The caller suggests to the protagonist that he must implement a transition from the fossil fuel economy to solar energy to save the world from ecological and social collapse. Following this call, he receives an email with the latest future models and technological advancements in solar technology to create a global transition. A few moments later, he receives another call from his future self in the year 2119. Solar energy has now dominated global markets, but the desired outcome does not prevail. The earth is now covered in silicon fields, as 'after all the deserts had been turned into solar fields, they began to devour arable land and vegetation cover' (Liu 2019, 64). The need for energy kept on growing and solar panels led to an excess of siliconisation. The caller proposes one more transition to save the future, ultra-deep drilling and a geoelectric form of green technology. Once again, the protagonist receives details on this advanced green technology and aims to implement it. However, another call takes place, and we find that the ultra-deep drilling did not solve the problems of the fossil fuel age. Instead, the earth surface is covered in radiation, an ever-expanding drilling force having destroyed the earth's magnetic field. The protagonist resolves to delete all the emails; having changed the course of history three times in one night, but in the end decided to change nothing at all. The result is an energy impasse not knowing how to move forwards and being trapped in the current capitalist system.

The tone of the narrative is one of technocratic authority, informed by the protagonist's belief that technology will provide the solution to a warming world. There is an emphasis on 'impossibility' where changes at a structural level are difficult to enforce, but human technological innovations are perhaps the answer to the existing problems of an impending energy crisis. As the protagonist's future self informs him, 'Remember I am speaking to you from the future. Think. We are smart people' (Liu 2019, 58). This entrepreneurial spirit and belief in humanity's progress recalls the techno-utopian discourse in energy politics explored by Imre Szeman. He notes that 'entrepreneurship exists in the twenty-first century as a commonsense way of navigating the inevitable, irreproachable, and apparently unchangeable reality of global capitalism' (Szeman 2015, 473). The apparent unchangeable reality of global capitalism is manifested in this narrative with disastrous consequences. The text primarily foregrounds the change in market demands and goals while providing the infrastructure necessary for wholesale global transition. Energy consumption continues at a fast pace with the introduction of solar power, as 'the need for energy kept on growing, and silicon plows had to devour more land' (Liu 2019, 65). The narrative appears to critique what Szeman describes as 'blind faith: the belief that some new technology will arrive to rescue us from our thorniest problems' (Szeman 2014). As Szeman further notes on techno-utopianism, 'this hardly constitutes a real solution to environmental and energy problems, which are produced by our way of life and not by bad technology' (Szeman 2014). In 'Moonlight', no transition is enforced in ways of living but simply the introduction of new technology. This is not to suggest technological innovation is not an important dimension of energy transition, but rather it cannot be solely relied on. As Szeman notes, his problem is not to do with technology itself:

> What I do have a problem with is our increasing tendency to look to technology and science to create a soft landing for capitalism. We are living in the first years of a fully capitalist world, which has led not to wealth for all but to gross economic inequalities across the planet. It has also produced an ever-hotter planet, with consequences that we are already beginning to experience, but which our grandchildren will have to endure. (Szeman 2014)

The demise of this techno-utopian belief system and entrepreneurial spirit is illustrated in Liu's narrative, as the proposed green and technological solutions lead to ever more ecological disasters. In the text, the solar industry is 'no different from the history of fossil fuel industries [...] land siliconization was even more damaging to the environment than desertification. As conditions deteriorated, drought swept the globe, and the occasional rainfall only resulted in massive floods' (Liu 2019, 65). The market growth and domination of solar forms of energy in 'Moonlight' are indicative of Szeman's critical statement that 'if we can't stop capitalism from destroying the planet, then we'll use technology to make the planet work with our consumer culture, whatever the ultimate cost might be' (Szeman 2014). Energy consumption is not changed or reduced but continues unimpeded at a mass scale as arable and vegetative lands turn into solar fields. For Andreas Malm, a liveable future requires renouncing all fossil fuels and 'living with the autonomous sun and wind and waves without any more solid energy to expand on' (Malm 2020, 228). Yet as Liu's narrative demonstrates, implementing renewables cannot be the sole response to climate catastrophe. Under the existing capitalist regime, whose path is a one-way system of progress, nature reveals its limitations. The fast growth of both solar and ultra-deep drilling and their ecological consequences in 'Moonlight' illustrate a need not to master and govern nature in order to meet consumer needs but rather to withdraw demand—to change Western norms introduced through the age of oil. Liu's narrative is of a circular nature; we find ourselves back at the very beginning where nothing has changed, and the fossil fuel economy remains. Such circularity reflects the logic of extractive capitalism and energy politics, imaginative and creative moves away from the fossil fuel industry are forestalled or prohibited as profit and the market is prioritised. As Wilson notes, 'the ontologies of modernity that have shaped the global present limit our ability to imagine other futures' (Wilson 2018, 381). Liu presents this crisis of capitalist modernity, with technological advancements providing only further issues and bringing us back to the dominance of fossil fuel economy.

Wilson expands on the current deadlock, suggesting that 'this energy impasse is the political, economic, and environmental deadlock created by the limits of Western ontologies and epistemologies that need to be newly thought' (Wilson 2018, 378). 'Moonlight' offers the outcome of these deadlocks and concludes with nothing happening at all, as 'the world began another ordinary day' (Liu 2019, 71). Capitalism remains the dominant socio-economic model and the fossil fuel industry the main source of energy to fuel it. The fantasy of techno solutions becomes apocalyptic in 'Moonlight', exposing how the problems of the present remain intact despite the implementation of renewable resources. Wilson's critique of Jonathan Porritt's book *The World We Made* highlights the flaws of foregrounding transition in primarily technological and economically driven solutions. It is, as Wilson suggests, 'business as usual disguised as radical innovation' (Wilson 2018, 380). The After Oil Research group discussion of solarities similarly illuminates the limitation of fossil fuel ideologies and how they may shape the move to renewables. As they contend:

> If solarities are to be just rather than unjust, they must be generated from below rather than from above; solar energy must be as dispersed as the sun's rays, refusing the kinds of concentration that petrocapitalism has engendered. (After Oil 2022, 40)

They suggest the limitations of capitalist top down approach to transition and technocratic solutions to the energy impasse. *Moonlight* likewise illuminates these failures and foresees the pitfalls of relying on technology alone.

The 'radical' innovations 'Moonlight' proposes to prevent ecological disasters are solely mediated through technology: a phone call, an email, presenting models of new and 'green' energy systems. Relations, and how we perform them, how we interact with our ecology, are significantly absent. This distant managerial approach provides the technology, yet not the cultural and social change to create a just and ethical energy transition. It could be argued that 'Moonlight' proposes inaction and convergence to the dominant energy system of the fossil fuel industry. However, its critique of violent energy transition illustrates the dangers which can ensue if we maintain the same knowledge systems and highlights the dominance of capitalist and techno-utopian solutions. For environmental change and justice to be carried out, Wilson suggests that 'from a feminist or Indigenous perspective, empathy, not just knowledge, clearly plays a role in action or stasis' (Wilson 2018, 383). Empathy is acutely absent from 'Moonlight'; the plot is driven by continuous phone calls, reports on climate and the introduction of new technologies. This is evident from the final phone call when the protagonist asks his future self 'Please tell me about our life after this moment' (Liu 2019, 69). The intimacies and personal life stories are entirely absent from the call; his future self refuses to reveal anything about their life. It is a narrative devoid of connection and interaction, solely driven by technological change and innovation. This lack of empathetic knowledge and engagement is perhaps one of the downfalls of their project to save the world from ecological destruction. 'Moonlight' looks to the limitations of technological solutions revealing what happens if we rely on the entrepreneur alone. Wilson's suggestion that empathy holds a key role in environmental action points to how affective relations are vital within the task of energy transition.

Wilson emphasises empathy in forming what she calls 'ecologies of care' (Wilson 2018, 379). Ecologies of care serve to counter the distant and managerial approach of 'Moonlight', introducing relations as an important dimension of energy transition. If, as Wilson contends, the project of oil capitalism and the nation state 'require as their prerequisite the erasure of certain bodies' (Wilson 2018, 396), the ambition of this chapter is then to illuminate these bodies—and particularly the nonhuman life forms erased from ethical consciousness. Like Wilson, who foregrounds empathetic knowledges from feminist and Indigenous perspectives, I too introduce 'Other knowledge' systems which prioritise encounters and relations between others. By introducing what Julietta Singh terms as 'dehumanist solidarities – social bonds that are mobilised and sustained through a refusal of sovereign human subject and enact agential forms of inhuman relationality' (Singh 2018, 123), we can find other ways of being and interacting with the world and its creatures. 'Moonlight' presents the violence of extraction and transition, but in what follows I look to the ways solidarity is formed across difference and how such a fellowship with animal or alien life forms can generate a different imaginary from the deadlocks of capitalist and techno-solutions to transition.

To return, however, to the question of SF and energy, Macdonald suggests,

> The eco-apocalypse genre usually moves in narrative arcs from violence to turpitude, depending on relative access to energy. The field is packed. But what of post-oil fictions of renewable energy? The utopian imaginations of a powered-down society seem less realized. (Macdonald 2016)

What I am about to discuss provides a response to Macdonald's query. Although not specifically about energy or post-oil worlds, *The Dispossessed* and *Solaris* open us to alternative relations beyond resource violence and human mastery, they expose the vulnerability of the human, bringing the human itself into question. In scenes of shared suffering, the undoing of mastery, alternative forms of living become possible in the novels, developing a sense of vulnerable co-existence and fellowship with other creatures, which has implications for energy transition.

Shared Suffering and Creaturely Companionship in *The Dispossessed*

Ursula Le Guin's *The Dispossessed* has been read 'as an anarchist, ecological, anti-capitalist, or revolutionary utopia' (Stillman and Davis 2005, xi). Le Guin's experiment in anarchism and anti-capitalist vision presents an antithesis to the failures of green capitalism we explored in 'Moonlight'. Despite the fact that energy transition is not the focus of Le Guin's fictional universe, the narrative illustrates an ecological sentiment which is resonant of Simone Weil's philosophical thought and Pick's creaturely poetics. An ecological philosophy emerges from *The Dispossessed*, a philosophy which is grounded in shared suffering and which offers a challenge to the regimes of extractive capital.

The Dispossessed is one of the novels from the Hanish cycle, alongside *The Left Hand of Darkness* and *The Word for World is Forest*. The Hanish cycle and other Le Guin fictions are often attuned to environmental politics, Indigenous rights and her own position as imbricated within settler colonial capitalism. In her essay, 'A Non-Euclidean View of California', she discusses how Utopia has always been white, European and male, consisting of one journey, the 'march of progress'. She goes onto note 'it seems that utopian imagination is trapped, like capitalism and industrialism and the human population, in a one-way future consisting only of growth' (Le Guin 2018, 111). Le Guin's experiment in an anarchist ambiguous utopia in *The Dispossessed* defies such capitalist logics, offering spaces of communion and solidarity through an environmental politics and a multispecies ethics.

The Dispossessed itself follows the twin planets Anarres and Urras, and the scientist Shevek as he ventures from Anarres, the anarchist planet, to Urras, a planet under a capitalist system, where the inhabitants are often described as propertarians. The Anarres people are settlers who left the political turmoil of Urras to build a society based on anarchist principles. The two planets have had little communication or relation to each other since the settlers arrived on Anarres 200 years ago. It is Shevek, the physicist, however, who changes this relation; his project, as he describes it, is 'to unbuild walls' (Le Guin 2014, 8). The narrative moves between stories from Shevek's past on the planet Anarres which involve resource struggles, episodes of drought and his ventures in physics to the present day where he navigates and attempts to understand the capitalist and hierarchical society of Urras with its constant wars, poverty and ideological divisions.

Scholarship on *The Dispossessed* focuses on the political tensions of anarchism and capitalist society, its relation to cold war politics, class conflict and its post-consumerist ethics. The intentions of my reading, although connected to Le Guin's anti-capitalist visions, are more specifically focused on the Weilian and creaturely aspects of *The Dispossessed* exploring what Pick terms as 'the communicative pressures of vulnerability' in the novel. The episodes of drought and famine on Anarres, Shevek's endurance of pain and suffering while carrying out his studies in physics and his intellectual musings on the significance of pain as a driving force for solidarity are my main foci. Another area that comes to the fore in my reading is the ecological aspects of the novel—Le Guin's focus on the wastefulness of Urras society, the ideas around possession, property and communal resources. As I explore these moments of isolated and shared suffering and the characters' sense of ecological awareness, I frame the analysis through Weil's philosophical thought, illustrating the ways in which my reading can gesture towards an alternative energy ethics from present extractivism.

Le Guin's *The Dispossessed* is neither strictly about energy nor nonhuman animals, yet the central character Shevek's fascination and experience of suffering and affliction provides an illustration of Pick's creaturely fellowship and Weilian philosophy. In my reading of *The Dispossessed*, I focus on Shevek's claim that 'it is our suffering which brings us together' (Le Guin 2014, 247). I thus demonstrate how the Weilian modality of affliction arises from Le Guin's novel offering what the After Oil collective strive for, regarding new ways to 'think, imagine, see and hear' (After Oil 2016, 41). Yoon Sook Cha suggests Weil's philosophy presents an other-centred ethics. By placing this other-centred ethics within the heart of energy transition, it can perhaps alter our relation to our wider ecology and other nonhuman life forms.

The Dispossessed is an experiment in anarchist principles, a world which Fredric Jameson has termed 'world reduction' (Jameson 2007, 267). Jameson focuses primarily on Le Guin's *The Left Hand of Darkness* but also draws on *The Dispossessed,* noting that Anarres is a planet where humans are entirely without biological partners. One reading of *The Dispossessed* is to see these two opposing worlds, Anarres and Urras, as an ambiguous utopian anarchism versus a dystopian rampant capitalism. Like Le Guin, Simone Weil was also a known anarchist exploring the inadequacy of Marxist tradition in her work *Oppression and Liberty* and favouring an anarchist approach to political change and justice. Although Le Guin is versed in anarchist politics, it is unknown whether she engaged with Weil's scholarship. However, Weil's philosophical and political writings, and Le Guin's fiction, have ecological themes that are foregrounded in their anti-capitalist visions. As Peter Stillman writes, 'Le Guin was writing at the outset of the modern environmentalist movement, symbolised by the first earth day in 1970' (Stillman and Davis 2005, 55). Weil herself never directly engaged with an environmentalist politics in her philosophical thought; however, contemporary scholars such as Simone Kotva and Anat Pick have identified the ways Weilian modalities are significant aspects in forming an ecological and creaturely consciousness. For Gib Prettyman, Le Guin is also a key figure in science fiction who has drawn on ecofeminism, Daoism and environmentalist politics. As Prettyman suggests, Le Guin's fictions work towards this 'recognition of ourselves in nature' (Prettyman 2014, 58). Bringing together Le Guin and Weil is a way of introducing a form of ecological thinking that has implications for energy transition. Weil's clear anti-capitalist thought and Le Guin's science fiction imaginaries are an attempt to unpack the violence of the present capitalist system, offering alternatives and different ways of being together.

For Gib Prettyman, 'ecology involves two related cognitive processes: unlearning the egoistic and anthropocentric illusions that underlie the psychic ecosystem of capitalism, and learning the real limits that characterise the material ecosystem and circumscribe human culture' (Prettyman 2014, 62). This definition of ecology is illustrative of both Weil's notion of decreation and Le Guin's fiction. As Sook Cha notes on Weilian decreative ethics, it is 'to begin to think of the way self-dispossession might be a passage to the other' (Cha 2017, 3). The self-dispossession that leads to a connection to another is also a passage of attunement to other life forms beyond species divisions. Eradicating the 'I' in Weilian philosophy becomes a way of recognising and hearing the call of the sufferer. Similarly, as Prettyman notes, Le Guin expresses an ecological strategy of 'trying to counteract the way of the ego' (Prettyman 2014, 65). The society of Anarres in *The Dispossessed* is led by the fictional ideology of Odonianism where egoism and individualistic possession is condemned and is replaced by a communal politics and philosophy. Le Guin's ecological strategy of counteracting the way of the ego in *The Dispossessed*, which is informed through her work on Daoism, is similar to Weil's scholarship on decreation. As Sook Cha describes, decreation is 'a radical decentering of the ego that would allow the other to make contact with others or, finally, with God' (Cha 2017, 17). Both Weil and Le Guin's focus on the decentring of the ego serves to implicate the human as earthly and ecological, as well as a spiritual being, connected and bound to others.

Weilian scholar Stuart Jesson describes Weil's concept of attention 'as a kind of openness or receptivity; a willingness to encounter – or even be penetrated – by what is given in the real' (Jesson 2017, 121). In this statement, Jesson describes how attention operates as a way of attuning oneself to suffering and the reality of our world. This receptive openness is illustrated by Le Guin's central character Shevek, and his desire to unbuild walls, to communicate across barriers—spatial, social and ideological. The image of the wall haunts Shevek from an early age. In the opening pages, the first sentence announces, 'there was a wall' (Le Guin 2014, 5). The wall, he describes, emphasised 'the idea of the boundary. But the idea was real' (Le Guin 2014, 5). From the very beginning, Le Guin accentuates both the idea and the materiality of walls and divisions, what separates and divides us from others. In this case, it is the separation of the two planets Urras and Anarres, but the complexity of these divisions is imbricated in the novel on a far deeper level from the cultural, ideological and the material. The walls which surround Anarres cut their universe off from communication with other planets, with Urras in particular appearing as alien, monstrous and monolithic. In his ambitions to create bridges between the two planets, Shevek is open and receptive in a Weilian sense, demonstrating a willingness to engage. In Shevek's dream as a child, he comes across a wall which 'went from horizon to horizon across the barren land' (Le Guin 2014, 30). He attempts to go on, but 'the wall stopped him' (Le Guin 2014, 30). The wall becomes a metaphorical and physical barricade to other worlds and fellowship with other creatures; the wall itself emphasises solidity and division, conflict and isolation. Shevek's ambitions throughout the narrative is thus to unbuild the walls which haunt him and surround his planet; his journey is one of openness and free communication beyond divisions and conflicts.

Weil's notion of attention allows for 'self-displacement, a loving patience, and a willing and receptive openness' (Jesson 2017, 126). Attention is an ability to wait, of emptying out oneself, in order to receive the other in all their unbearable reality and truth. Weil discusses this at length in *Reflections on the Right Use of School Studies.* As Weil writes, 'above all our thought should be empty, waiting, not seeking anything, but ready to receive in its naked truth the object which is to penetrate it' (Weil 2009, 4). We find in Shevek's character this process of opening oneself out. The narrator describes how

> Shevek had learned how to wait. He was good at it, an expert. He had first learned the skill of waiting for his mother Rulag to come back, though that was so long ago he did not remember it; and he had perfected it waiting for his turn, waiting to share, waiting for a share. At the age of eight he asked why and how and what if, but he seldom asked when. (Le Guin 2014, 39)

It is this practice of waiting that allows one to hear and attend to another's suffering. As Weil suggests, 'the love of thy neighbour in all its fullness simply means being able to say to him: what are you going through' (Weil 2009, 124). The ability to patiently wait is a process of allowing to hear what is often not articulated or is sometimes ignored. Shevek's method of waiting provides an avenue to his ambitions of breaking down walls, to communicate and listen across divisions.

Weil's concept of decreation explores this sense of receptive openness. Decreative ethics questions the sovereign subject and thus offers an undoing of power relations. As Sook Cha notes, 'to not exercise power where one has it, then, is so radical because it withdraws altogether from the binary model of subjugator subjugated' (Sook Cha 2017, 4). In such a process of radical dispossession, decreative ethics leads to a loss of oneself, an inability to say 'I'. The conceptions of individualism and autonomy that shape capitalist ideology are thus brought into question through this dispossession. It becomes not so much a question of rights and individual freedom but of obligations and willing attention. We find this dispossession of the 'I' in Le Guin's anarchist world Anarres through the political project of Odonianism. As a young boy growing up, Shevek insists on being in the sun. He enjoys the radiating heat and declares 'Mine! [...] Mine sun!' (Le Guin 2014, 26). Following this, his father insists that 'Nothing is yours. It is to use. It is to share. If you will not share it you cannot use it' (Le Guin 2014, 26). As Shevek is reduced to tears, his father continues: 'Come on, you know you can't have things' (Le Guin 2014, 26). In Anarres, ownership and property are condemned, and a communal philosophy and practice is instead favoured. To declare ownership, to focus on what 'I' am or what 'I' own is what the planet's inhabitants' term egoising. By forging such an ethics and politics around a dispossession of the 'I' in favour of a communal and shared subjectivity, Le Guin's anarchist world in Anarres follows that of Pick's creaturely ethics which is inspired by Weilian thought. As Pick concludes in her work, 'a creaturely ethics [...] does not ask what are the limits of rights? But what are the limits of attention?' (Pick 2011, 193). Weil's philosophical thought argued for the precedence of obligations over rights—rights being tied to property and ownership, whereas obligations emerge regardless of individual freedoms and autonomy. In Weilian thought, it is only through sincere and bound obligations in which we can hear the call of the suffering that we can become aware of what someone is going through and attend to their cry. As Sook Cha questions,

> What kind of ethical perspective—freed from the egocentrism of "the person"—can be developed from the kind of impersonal being Weil envisages? Would we find there a distinctly ethical expression of being, one that respects the alterity of the other so that it can neither be subsumed under nor mastered through one's responsibility toward him? Would it compel us toward a mode of being profoundly open to the other, not because of any supposed social harmony (which can lead to idealization, in Weil's view) or because the relationship is reciprocal (obligation, according to Weil, does not belong to an economy of equitable exchange or reciprocity), but because both the self and the other are, finally, vulnerable to harm? (Cha 2017, 20)

Le Guin, like Weil, attempts to envisage a world in which one is freed from egocentrism and engages instead within a reciprocal relation, forging a path to undoing sovereignty and ownership. It is this question of vulnerability, of both being subject and vulnerable to harm, in which Le Guin's character Shevek suggests brotherhood is formed. As he announces to his friends when discussing the severity of pain, 'I'm trying to say what I think brotherhood really is. It begins – it begins in shared pain' (Le Guin 2014, 54). Shevek illustrates how solidarity and companionship emerge from both being vulnerable to harm, subject to blind force and gravity in the Weilian sense. Shared vulnerability can extend not just to the human being but to other creatures and ecological life forms. If solidarity across difference begins with pain, what implications does this have for our treatment of nonhuman life forms? Can a recognition of shared pain lead to a fellowship with other beings that are similarly finite and temporal, and vulnerable to harm? As the title of Le Guin's novel is indicative of an act of dispossession, of becoming the dispossessed, Weil's decreative ethics is thus illustrated through Le Guin's renunciation of the 'I' and instead proposes a creaturely companionship on the basis of shared pain and suffering.

During Shevek's studies, he contemplates the experience of pain, its purpose, and its implications. As he notes, pain

> Exists [...] It's real [...] Suffering is the condition on which we live. And when it comes you know it. You know it as the truth. Of course it is right to cure diseases, to prevent hunger and injustice, as the social organism does. But no society can change the nature of existence. We can't prevent suffering. This pain and that pain, yes but not Pain. A society can only relieve social suffering. The rest remains. The root, the reality. All of us here are going to know grief; if we live fifty years, we'll know pain for fifty years. And in the end we'll die. That's the condition we're born on. (Le Guin 2014, 53)

Weil's philosophy is illustrated in this passage, describing a case in which reality is made 'not less painful, only clearer' (Weil 2021, 1) and where vulnerability itself is 'a mark of existence' (Weil 2002, 108). As Shevek contemplates a past experience of caring for another, a man who crashed an air car, the vehicle having caught fire, Shevek suggests: 'You couldn't do anything for him, except just stay there, be with him' (Le Guin 2014, 54). It is the act of bearing witness to another's pain, hearing their cry, which is where solidarity and companionship is formed. As Sook Cha suggests,

> To recognize this vulnerability implies a failure, a failure that is even beautiful, but beautiful in that it cannot defend itself against the cry of the other. That is the ethical bind: that, bound to the other by an unconditional and unconditioned demand ("Why am I being harmed?"), one must nonetheless undergo and sustain an unconditioned desire defined by its unrealization and that presumes the destruction of one's "I." (Cha 2017, 39)

As Shevek hears the cry of the man he attends to, he realises he has little to offer, but in hearing the suffering of another, this connection enables an ethical bind in which one's self becomes undone in the face of an other's pain. Much later in the novel, as Shevek joins a protest on the planet Urras, he announces to the crowd: 'It is our suffering that brings us together [...] the bond that binds us is beyond choice' (Le Guin 2014, 247). This ethical bind, formed through the dispossession of the 'I', has implications for a new and alternative relation to energy. On Anarres, they use wind power as their main source of energy. Unlike the mistakes and violent transitions in 'Moonlight', Le Guin presents a world of de-growth and decreation on Anarres—a world focused not on possession, property ownership and extraction but formed through solidarity, mutual aid and community.

The energy impasse is 'without signposts', and we hold the immanent knowledge that 'we now must transition to different ways of being in the world, both with each other and in relationship to the environment' (After Oil 2016, 17). Le Guin's expression of a decreative and other-centred ethics, however, can gesture towards a changed relation to the environment and energy sources. As I have now demonstrated the Weillian aspects of *The Dispossessed*, my reading suggests that an other-centred and creaturely ethics can be orientated towards ecological consciousness. 'Upending sovereign modes of identity and political constitution' (Cha 2017, 6) allows for our vulnerable co-existence to become apparent, an earthly yet spiritual bind to the cries which often remain muted or silent.

Le Guin's Anarres is not, however, a strictly utopian world, with its frequent famines, resource depletion and, as Jameson has noted, the absence of biological partners, namely fellow creatures. Anarres is a barren landscape prone to droughts and inhabitable for other species. Shevek is astounded when arriving on Urras at the flora and fauna that populate their world:

> It's a queer situation, biologically speaking. We Anarresti are unnaturally isolated. On the old World there are eighteen phyla of land animal; there are classes, like the insects, that have so many species they've never been able to count them, and some of these species have populations of billions. Think of it: everywhere you looked animals, other creatures, sharing the earth and air with you. You'd feel so much more a part. (Le Guin 2014, 186)

This feeling of integration with different species in Urras emphasises becoming one creature among many, a fellowship emerging as you become a part of an entire ecosystem. This sense of belonging, however, emerges not just with Shevek's fascination on the planet Urras but also on Anarres with his partner Takver and her role as a biologist. As Shevek describes,

> Her concern with landscapes and living creatures was passionate. This concern, feebly called "love of nature", seemed to Shevek something broader than love. There are souls, he thought, whose umbilicus has never been cut. They never got weaned from the universe. They do not understand death as an enemy; they look forward to rotting and turning into humus. It was strange to see Takver take a leaf into her hand, or even a rock. She became an extension of it: it of her. (Le Guin 2014, 154)

The umbilical cord image envisions how Takver is bound to her world, the universe and the creatures which inhabit it. She is depicted as an extension of her environment, rather than a solitary master of it. By becoming an extension to the rock or the leaf, Takver is not the sovereign subject or master but exists alongside these organic and elemental forms. Le Guin's description envisions a creaturely companionship, the human itself immersed and connected to other life forms, a process of 'retranslating the human into the creaturely order, into the anonymity of perishable matter' (Pick 2011, 183). Le Guin illustrates the unbuilding of walls to uncover long-forgotten bonds and ties, the connections between the two alien worlds and the creatures which reside in them. Le Guin thus envisions Pick's creaturely project of 'illuminating the relations we currently have and the ones I believe we ought to have – with the world around us' (Pick 2011, 185).

The Dispossessed concludes with Shevek releasing his new theory of physics, thus making possible a model of communication that can traverse the universe to speak with other planets from different time zones and locations. He does not do this for entrepreneurial gain or to benefit the profit seekers from Urras but to unbuild the walls, to walk through them and to connect with others. What strikes Shevek as the connection that binds us all, however, is not the connections formed through scientific innovations but through the 'deepest intimacy of pain' (Le Guin 2014, 304). As with Weilian philosophy, Le Guin forges a world and a sense of solidarity through our vulnerability. In this vision, 'man as the centre of the universe is no more, and a new history, a musical and natural history – the natural history of creatures – is born' (Pick 2011, 183). Le Guin's fictional accounts of ecological relationships lead us to 'the ordinary – yet extraordinary – living encounter' (Pick 2013, 13) with vulnerability. Suffering becomes the source of solidarity in *The Dispossessed*; in Julietta Singh's terms the novel 'asks us to open ourselves to reimagining ways of relating to each other – to others, human, nonhuman, and inhuman, to which (even when disavowed) we are mutually bound' (Singh 2018, 3). The ethical bind that ties us to other beings—human and nonhuman—is distinctively non-masterful, allowing for a fragile opening and passage to another. It is through this shared vulnerability that one can imagine a world beyond violent modes of extraction, an ethical and non-anthropocentric project that 'is a profoundly hopeful one that gazes towards a future it still cannot see' (Singh 2018, 21).

For the After Oil Research collective, 'the struggle that is currently taking place over the direction of energy transition, which involves scientists, activists, governments, and businesspeople, is a struggle over representation and narrative, the stories we tell about human capacity and future possibility' (After Oil 2016, 73). Perhaps it is the undoing of certain human capacities, exploring our limits and finitude—seeing ourselves as perishable matter that enables the blockades and walls preventing a just and ethical energy transition to become unbuilt. Le Guin's *The Dispossessed* and Weilian philosophy provide renewed and alternative modalities of the human through the experience of suffering and affliction. Instead of being a subject holding dominion and mastery over the earth, the human becomes one creature among many, vulnerable and subject to pain. Their reconfigurations of the human in fiction and philosophy offer an entryway to alternate modes of understanding and interacting with energy. Thus, the combined analysis of Le Guin's fiction and Weil's philosophy facilitates engagement with the violence of the human past and present yet envisioning a future in which we become attentive to different ways of hearing, seeing and living upon the earth amongst its different creatures, life forms and energies.

Undoing Systems of Mastery in *Solaris*

Ursula Le Guin's 'Foreword' to the Russian SF writers, Arkady and Boris Strugatsky's *Roadside Picnic*, offers an insight into Stanislaw Lem's *Solaris* and the failures of human mastery. Le Guin describes that:

> The question of whether human beings are or will be able to understand any and all information we receive from the universe is one that most science fiction, riding on the heady tide of scientism, used to answer with an unquestioning Yes. The Polish novelist Stanislaw Lem called it "the myth of our cognitive universalism." *Solaris* is the best known of his books on this theme, in which human characters are defeated, humbled by their failure to comprehend alien messages or artifacts. They have failed the test. (Le Guin 2012, viii)

In the novel *Solaris* I explore the implications of what it means to 'have failed the test' and how the undoing and failure of human mastery leads to a discovery of different relations beyond the human. Mastery shapes our past and present relation to energy sources—from the whale in *Moby-Dick*, hunted and harpooned, to animal bodies, processed and packaged, from oil itself extracted and refined. These power relations, that have come to shape energy systems and politics, rely on the human holding both dominion and sovereignty over the natural world.

Lem's *Solaris* opens with the SF endeavour of discovery and mastery, yet as the novel develops mastery begins to falter and new relations emerge. The undoing and failure of mastery are now my focus, exposing how mastery's failure leads to the discovery of both vulnerability and 'non-masterful forms of politics' (Singh 2018, 15). The planet Solaris is made solely of water, a nonhuman agent that refuses to be controlled or harnessed by human scientific and technological practices. Just like the whale in *Moby Dick*, the planet is surveyed and assessed according to human measurements while being subjected to violence through both classificatory language and material modes of extraction. In the narrative, there is a transition from distant scientific observer, the master of its object of study, to 'a participant in mutual exploration' (Jue 2014, 231), which is the result of mastery's failure. This failure calls for new imaginative horizons of how we engage and live with other nonhuman life forms, thus problematising our current violent practices of energy extraction. In my exploration of *Solaris*, I turn to the ruptures of mastery and control, the moments of vulnerability which lead to forms of creaturely companionship.

Capitalist forms of production and extractive modes of viewing are once again at the heart of my critique of *Solaris*. I suggest Lem's critique of human mastery offers a challenge to the current capitalist imperial system despite emerging from a location occupied by the Soviets and a communist context. I suggest that the novel's challenge to hegemonic thinking and violent regimes can offer a critique and alternative to capitalist logic unravelling mastery in all its forms.

Julietta Singh's project in *Unthinking Mastery* follows a similar trajectory to Pick's claim in her work which is to make, in part, dehumanisation something radically positive. Singh's project differs somewhat from other posthuman scholars in the sense it is committed to critical conversations with the decolonial. Singh's dehumanism is a project which aims 'to become exiled from subjectivities founded on and through mastery [...] a project of remaking the human from the outside' (Singh 2018, 5). Addressing both the animality of the human, the ideological fantasies of human's unique agency, Singh aims to create other less masterful subjectivities that are driven by the promises of vulnerability. Like Weil and Pick,

Singh also emphasises vulnerability as the modality that transforms not only our understanding of the human identity but also their wider relations. As Singh argues,

> If the masterful work of global imperialism functions through dehumanisation of those it aims to conquer, and if we can now argue that the human to which we have been aspiring is intimately bound to a logic of mastery, the looking toward those "other genres of being human" that have been well lived and will be lived by those subjected to imperial force might offer us other performances of the human that allow us to begin to practice a non-masterful politics. (Singh 2018, 15)

Looking to these other genres of being human can provide a way of becoming attuned to other species and the environment. It is through the colonial SF endeavour within *Solaris* in which I explore the human's inability to master and its ramifications for energy relations and transition. Science fiction's ability to 'provide us with a future-orientated perspective' (Vint 2010, 21) might offer ways of changing the here and now, gesturing towards a renewed relation to energy.

Solaris provides an acute illustration of the limitations of human mastery and discovery, opening us to scenes of vulnerability. As Melody Jue suggests, the central crisis of the novel is 'that human beings can only know what is Other through existing frameworks of cognition and linguistic means' (Jue 2014, 230). *Solaris* is a narrative that begins as a colonial venture, the scientific community's aim of understanding and thus mastering the planet. The crisis of human mastery is demonstrated by the scientific project called Solaris studies, which is the academic pursuit of understanding the mechanisms of this ocean-based planet. Theorised as a 'thinking ocean', the scientists produced multiple hypotheses of its creative movements, attempting to understand the planet's processes and interactions through mathematical models and equations. The project evoked all kinds of metaphysical questions: 'Was it possible for thought to exist without consciousness? Could one, in any case, apply the word thought to processes observed in the ocean?' (Lem 2003, 25). For Melody Jue, these ruptures in understanding illustrate the novel's central theme: 'a crisis that is jointly scientific, masculine, colonial, and terrestrial' (Jue 2014, 227). These crises are in effect the failure of mastery, and therefore, as Singh suggests, 'in failing to master [...] we become vulnerable to other possibilities of living, for being together in common' (Singh 2018, 21). It is to this process of being together in common that the characters of the novel find themselves unable to master or understand the planet, they become open to different and perhaps less masterful forms of relations.

The novel follows Kelvin, a psychologist, who ventures out to station Solaris to join a team of scientists who are hoping to understand the ocean depths of the planet. But as Jue notes, 'Lem's fantastic ocean resists both physical and epistemic human penetration with depths that remain cognitively out of reach to whatever extent they exists at all' (Jue 2014, 228). As Kelvin arrives on station Solaris, he notices something strange about the other scientist onboard, both Snow and Sartorius. Snow is elusive towards him, concerned and uncertain about whether he is a fellow human being. Sartorius locks himself in his room to hide away from Kelvin. The sense of apprehension onboard the intergalactic station suggests to Kelvin that something unnerving has taken place. Gibarian, a colleague from earth who was also placed on the station, appears to be absent. Kelvin interrogates Snow about Gibarian's disappearance only to discover that Gibarian has died in mysterious circumstances. Later Kelvin discovers the cause behind this apparent madness, as he is visited by one of the ocean planet's avatars, the physical manifestation of his memory of a woman he once loved. The planet, according to Snow and Gibarian's notes, is sending them their past memories of people who were once a part of their lives. As if reading the scientists' dark and forgotten memories, the planet creates clones of their past lovers, children and mothers, following and haunting the scientists' every movement onboard the station. However, rather than reading these human clones through a psychoanalytic frame, as pure manifestations of the character's unconscious, I look to the implications of them as material nonhuman life forms, an encounter with the alien Other.

For Ann Weinstone, the subject of *Solaris* is 'colonisation, especially those colonizing activities that proceed from hegemonic science' (Weinstone 1994, 177). The scientific observer combined with the colonial mastering gaze resonates with introductory exploration of the whale in *Moby-Dick*, as being subject to over-coding and mastery through scientific naming and classifications. In the novel, the planet Solaris becomes the object of human progress and innovation as the scientific community's primary aim is to make 'contact'—an aim that is arguably colonial, conjuring images of the final frontier. Solaris itself becomes interpreted in many ways: a colossal brain, a thinking ocean, a mass energy source, and a strange and alien animal.

Much of the novel explores the scientific reports, investigations and experiments carried out on the planet Solaris, as Kelvin attempts to decipher what is happening to them onboard the station. In the chapter Solarists, Kelvin surveys the history of the scientific community's attempts to make 'contact' with the ocean. First attempts at contact were developed through an electronic apparatus: 'the ocean itself took an active part in these operations remodelling the instruments. All of this, however, remained somewhat obscure. What exactly did the ocean's participation consist of?' (Lem 2003, 21). As Kelvin further notes, 'constantly it seemed, the experts were on the brink of deciphering the ever-growing mass of information' (Lem 2003, 22). Initially, the scientists suggest the ocean expressed itself 'in a more or less mathematical language' (Lem 2003, 22). Lem's focus on the scientific endeavours, their successes and, ultimately, their failures to make contact, in order to understand this nonhuman and alien agent, resonates with ecofeminist Vandana Shiva's critique of modern science. As Shiva notes in her and Maria Mies' key work, *Ecofeminism*:

> Modern science is projected as a universal, value-free system of knowledge, which by the logic of its method claims to arrive at objective conclusions about life, the universe and almost everything. This dominant stream of modern science, the reductionist or mechanical paradigm, is a specific projection of Western man that originated during the fifteenth and seventeenth centuries as the much acclaimed Scientific Revolution. (Shiva et al. 2022, 22)

In this work, Shiva addresses how often modern scientific discourses can become reductionist, stripping nature to its bare mechanics and thus presenting the earth as inert and fragmented matter. Under such a reductionist paradigm, nature is understood as mere commodity, thus lending itself to a capitalist framework, the environment seen as something to be harnessed, controlled and conquered. The study of Solaris is similarly an Enlightenment project, demonstrating the ultimate progress of 'man' as well as being tied to capitalist progress. This is evident through the scientists offering a prize for those who could discover ways of harnessing the energy of the ocean planet and transporting it back to earth. As Kelvin notes:

> Two years before I began the stint in Gibrarian's laboratory which ended when I obtained the diploma of the Institute, Mett-Irving Foundation offered a huge prize to anybody who could find a viable method of tapping the energy of the ocean. The idea was not a new one. Several cargoes of the

> plasmatic jelly had been shipped back to Earth in the past, and various methods of preservation had been patiently tested: high and low temperatures, artificial micro-atmospheres and microclimates, and prolonged irradiation [...] the end product was always a light metallic ash. (Lem 2003, 177)

The attempts to extract and transport the energy of Solaris fail after multiple attempts at preservation and transportation to Earth. This alien planet refuses to be harnessed for capitalist innovations; human attempts at control and extraction are met with failure. The conflation between scientific endeavours and capitalist profit demonstrates Shiva's argument in *Ecofeminism* (Shiva et al. 2022) where she explains the reductionist perspective as serving the dominant economic and political systems. As Shiva suggests, 'the reductionist world-view, the industrial revolution and the capitalist economy are the philosophical, technological and economic components of the same process' (Shiva et al. 2022, 24). Lem thus demonstrates in *Solaris* how scientific mastery is a capitalist imperial motivation. The depiction of the planet as an energy resource here follows an extractivist logic, in which 'we associate energy with something to be burned, hoarded, or wasted without any clear end' (Marder 2017, x). However, as Kelvin's stay on station Solaris continues, we find the planet is not simply inert matter but has some kind of agenda of its own. The inability to extract and refine the energy of Solaris holds significant implications for energy transition, illustrating the undoing of mastery results in a withdrawal.

Before any form of withdrawal of power takes place, however, we witness the scientists struggle with this nonhuman agent. For Melody Jue, the critical problem *Solaris* stages 'is the cul-de-sac of scientific investigation that brackets the observer out of the dynamic relation between phenomena/other and self' (Jue 2014, 230). This bracket creates a subject/object dichotomy in which the planet is stripped of any form of agency, the planet is thus shaped under the colonial and scientific gaze. As Snow speaks with Kelvin on their scientific intentions, we discover the legacies of colonial power beneath their pursuit of contact. As Snow suggests:

> We don't want to conquer the cosmos, we simply want to extend the boundaries of Earth to the frontiers of the cosmos. For us, such and such planet is as arid as the Sahara, another as frozen as the North Pole, yet another as lush as the Amazon basin. We are humanitarian and chivalrous; we don't want to enslave other races, we simply want to bequeath them our values and take over their heritage in exchange. We think of ourselves as the Knights of holy

> Contact. This is another lie. We are only seeking Man. We don't know what to do with other worlds. We need mirrors [...] We are searching for an ideal image of our own world: we go in quest of a planet, of a civilisation superior to our own but developed on the basis of a prototype of our primeval past. (Lem 2003, 76)

This passage evokes the colonial venture of appropriation and assimilation, the search for and discovery of new lands and planets as a means to enforce the image of the Western man as the measure of all things. The description of themselves as 'humanitarians' is not to be interpreted as demonstrating acts of humility and generosity but can be aligned with Sylvia Wynter's concept of the Colonaility of Being. For Wynter, the human is often defined according to the Western bourgeois conception of what is and what is not human. She then suggests that 'the negation of the generic "normal humanness," ostensibly expressed by and embodied in the peoples of the West' (Wynter 2003, 266). This normativity is often reflected in Lem's *Solaris* as the scientists grapple with what is distinctively Other than themselves. As Wynter goes on to note in reference to the treatment of America's Indigenous populations, there is an 'overrepresentation of Man as if it were the human, and to legitimate the subordination of the world' (Wynter 2003, 267). Wynter's depiction of the coloniality of 'man' echoes in Lem's narrative as the scientists attempt to subordinate the nonhuman life forms according to their measurement of what is human. If we are instead to follow Singh's practice of dehumanism, to introduce different genres and modalities of the human, something that might not appear to be human at all, this formulation of the Coloniality of Being must be undone. *Solaris* captures Singh's notion of dehumanist practice by exposing vulnerability and the undoing of scientific mastery in the face of a nonhuman creature who refuses to become absorbed or assimilated under the colonial gaze.

In *Energy Dreams,* Michael Marder notes that 'of Greek provenance, the word energy is stamped by a double entendre. Composed of the prefix en- and the noun ergon, energia can be literally translated as "enworkment", putting-to-work, activation' (Marder 2017, 3). Such analysis of the term 'energy' suggests that it is associated with activation, an implicitly extractive logic in which the natural resource is put to work or activated. *Solaris* begins with this dominant form of logic and with the scientists' ambitions of activation, contact and extraction. However, as we witness the arrival of what the scientists term phi-creatures, the material manifestations of their memories, the notion that the planet Solaris is inert and unthinking matter is brought into question.

Singh's *Unthinking Mastery* is illustrated by Kelvin's narrative journey within *Solaris*—a process of moving from human mastery to vulnerable co-existence with alien life forms. Following a similar theoretical arc to those of Weil and Pick, Singh orientates us to vulnerable modes of being. As she suggests, 'the practice of vulnerable reading can move us "beyond" mastery not in the sense of exceeding it but in the sense of surviving it in order to envision being in and for the world' (Singh 2018, 23). After the violence of attempting to harness, grasp and understand the ocean planet, the novel opens us to this form of vulnerable reading; no longer the distant scientific observer, Kelvin becomes immersed in a co-relation with the planet and its phi-creatures. His vulnerability is exposed through his relation to Rheya, the ocean planet visitor. Initially, Kelvin is shocked and terrified by the presence of his dead wife now physically materialised through the power of the ocean planet. He attempts all means of murderous aims in order to rid himself of her. Kelvin manages to escape her by placing Rheya in a launch rocket. In this separation, she screams, he declares, 'I could not bear to expose myself again to the sound of that horrifying voice, which was no longer even remotely human' (Lem 2003, 68). Kelvin conceives Rheya as distinctly Other, outside the category of the human. His logic of dehumanisation serves as a way of separating himself from Rheya and the planet Solaris itself, simultaneously asserting a form of power and protection from what is Other.

Rheya poses a crisis of understanding for Kelvin; she has often been read as a manifestation of Kelvin's unconscious, revealing his dark repressed memories. However, Rheya is autonomous from both the planet's intentions and Kelvin's memory of her. It is perhaps more convincing to read Rheya as a material creature. Kelvin's encounter with Rheya is therefore not a confrontation with his inner psyche but rather the encounter with the materiality of an alien creature. In Weinstone's terms, 'Rheya resists oversignification by disrupting totalizing categories such as subject/object, human/nonhuman, and biological entity/machine. As a result, the scientists are forced to confront the limitations of their unidirectional world view and must grapple with the possibility that they inhabit a world of multiple, constitutive, and sometimes unalterably alien agencies' (Weinstone 1994, 177). As Rheya reappears after Kelvin's first attempted escape from her, the dynamic between them alters. He discovers from Gibraian's recordings that she is not in fact Rheya, but what Gibrarian understands to be some sort of instrument or appendage from the ocean planet. As Weinstone further notes, 'Rheya stands in a unique ontological position. She is not one of the freestanding beings, i.e., she is neither one of the humans nor the alien Ocean. She occupies a gap, brought into existence only to serve as a parodic critique and perhaps as a bridge between two master signifiers: she is the oversignified creation par excellence' (Weinstone 1994, 179). Rheya's fluid identity of existing within the in-between, as a bridge between the human and the alien, brings into question the very nature of the human itself. She can feel pain, is a vulnerable being, but her mortality is not entirely clear. As Sartorius suggests, 'The Phi-creatures reappear exactly as they were, down to the last detail ... as vulnerable as before, each time we attempt to rid ourselves of them' (Lem 2003, 108). The key term here is the definition of the Phi-creatures as vulnerable beings—subject to pain and force. Despite Rheya's unique difference from human life, she, like all living beings, is vulnerable. To recall Simone Weil, she perceives vulnerability as a marker of existence. It is thus through such revelations of vulnerability that the bridge between the human and the alien is crossed.

This recognition of mutual vulnerability does not inevitably lead to an ethical response. As Kelvin discovers through his reading of scientific manuals, he finds that 'following the Eruption of the 106, and for the first time in Solarist studies, there were petitions demanding a thermo-nuclear attack on the ocean. Such a response would have been more cruelty than revenge, since it would have meant destroying what we did not understand' (Lem 2003, 130). The desire to attack Solaris with nuclear weapons comes from a motivation to destroy what they cannot access, know or control. As with the Phi-creatures, and Rheya in particular, the scientists onboard the station want to rid themselves of the creature they cannot understand or control. Kelvin alternatively wishes to maintain intimacy and connection with Solaris and Rheya. His recognition of their corporeality and vulnerability creates a change in Kelvin. Such transitions of becoming open to what is animal, alien or Other provide an answer to Alexander Weheliye's question: 'what different modalities of the human come to light if we do not take the liberal humanist figure of Man as the master-subject but focus on how humanity has been imagined and lived by those subjects excluded from this domain' (Weheliye 2014, 321)? Rheya thus occupies such a realm for the scientists, the space of the excluded—an alien life form which if it is not mastered, must be expelled or destroyed.

Solaris explores the domain of the excluded, illustrating the agency of the alien life forms which are beyond our grasp.

Kelvin's process of withdrawing mastery, of recognising the beauty of the planet ocean Solaris, without choosing to conquer or invade, but to engage in a mutual exploration, opens us to Pick's concept of 'letting be'. Pick engages with Laura Mulvey's notion of the cinematic male gaze, a look which at once devours and masters. She presents the antithesis of the male gaze through Weilian philosophy that inspires her concept of letting be. As Pick suggests, 'Thus, beauty involves a "letting be" of the object before us. In place of the hungry gaze, Weil makes possible a non-devouring mode of looking, a "vegan gaze" cast by the camera, viewer, or critic. While the devouring look yields pleasure, looking without devouring is akin to love' (Pick 2018, 128). The vegan gaze, for Pick, offers distance and withdrawal yet an intimacy akin to love. We witness this transition in Kelvin as he chooses to leave station Solaris and visit the planet itself. As he describes,

> Alone over the ocean, I saw it with a different eye. I was flying quite low, at about a hundred feet, and for the first time I felt a sensation often described by explorers but which I have never noticed from the height of the Station: the alternating motion of the gleaming waves was not at all like the undulation of the sea or the billowing of clouds. It was like a crawling skin of an animal – the incessant slow-motion contractions of muscular flesh secreting a crimson foam. (Lem 2003, 210)

It is particularly significant that Kelvin emphasises viewing the ocean with a different eye. This transformation recalls the After Oil Research Collective's demand to discover different ways to see and hear. Kelvin no longer clings to scientific mastery in order to control the object of study but enters into a co-relation of a mutual explorer. Kelvin's description of the planet also emphasises the corporeality of the planet, its contours and movements compared to the skin of an animal. His encounter with the ocean is mysterious and withdrawn, yet they appear as both open and receptive to one another. The process of withdrawing, a sense of reticence, allows Kelvin to engage in a form of openness. For Pick, letting be is 'a conservationist impulse that honours the existence of beings and things by looking-without-devouring' (Pick 2017, 47). Unlike the scientific and colonial gaze to which the planet has been subjected, Kelvin follows Pick's conception of looking without devouring, engaging in a non-masterful politics.

As Kelvin explores deeper into the ocean planet, he realises 'that [he] was not in the least concerned with the mimoid but to acquaint [himself] with the ocean' (Lem 2003, 212). Rather than attempting to extract the energy from the planet surface, to understand its intentions and mechanisms through scientific innovations, Kelvin instead wishes to address and recognise the planet as a fellow creature and life form. The interaction between the ocean planet and Kelvin is one of curiosity and playfulness:

> I raised my hand slowly, and the wave, or rather an outcrop of the wave, rose at the same time, enfolding my hand in translucent cyst with greenish reflections. I stood up, so as to raise my hand still higher, and the gelatinous substance stretched like a rope, but did not break. The main body of the wave remained motionless on the shore, surrounding my feet without touching them, like a strange beast patiently waiting for the experiment to finish. (Lem 2003, 212)

The two engage in a form of mutual exploration, what Melody Jue describes as 'the co-creation of meaning practiced by two aware participations in mutual curiosity' (Jue 2014, 238). As Jue goes on to note, science fiction stories such as *Solaris* 'ask us to rethink the way that we position ourselves in relation to the waters, to others, and how the dynamic tidal space of contact might offer alternate and more mutualistic space' (Jue 2014, 239). It is to this mutualistic space which in the *Dispossessed* was discovered through shared affliction, which Kelvin discovers through becoming open and vulnerable to another. As Kelvin concludes: 'We all know that we are material creatures, subject to the laws of physiology and physics, and not even the power of all our feelings combined can defeat those laws' (Lem 2003, 213). This recognition of himself and others as both material and temporal beings recalls Pick's creaturely ethics that is driven through the logic of flesh. As Pick notes, 'Human (and other bodies) are indiscriminately subject to natural necessity and powers from without' (Pick 2011, 4). To push Pick's creaturely poetics further, so too are energy sources subject to mastery, necessity and the forces that penetrate them from the outside. By recognising vulnerability as a modality that connects us all from other species to plant and organic life, we find solidarity and fellowship with the most unlikely of creatures and things.

Marder's *Energy Dreams* looks towards a non-extractive paradigm to understand and interact with energy. By exploring the failures of extraction and mastery to a mutual exploration and shared vulnerability in *Solaris*, I too have proposed a non-extractive paradigm that is simultaneously a creaturely ambition and a non-masterful one. *Solaris* exposes the inhuman solidarities that can be formed across difference and presents ways of reconceiving energy—to not aimlessly devour and consume but to withdraw and acknowledge the beauty of difference, a process of letting be. Science fiction immerses us in worlds unfamiliar and alien, where we encounter animal-like creatures, unknown and unmastered; the texts demonstrate alternative ways of understanding the here and now in order to map a just and ethical future.

The Energy Impasse

Adrienne Maree Brown's *Emergent Strategy, Shaping Change, Changing Worlds* focuses on science fiction as a world-building process. As she declares, 'I would call our work to change the world "science fictional behaviour"—being concerned with the way our actions and beliefs now, today, will shape the future, tomorrow, the next generations. We are excited by what we can create, we believe it is possible to create the next world' (Brown 2017, 14). Science fiction, as it is conceived by Brown, is rooted in a decolonial and feminist practice, a way of practising a just future together, and opening up new and alternative relations to others and the world. 'Moonlight', *The Dispossessed* and *Solaris* explore scenes of energy transition, moments of mastery and violent extraction, encounters with alien and animal Others, and the implications of the suffering of another. The novels experiment, critique and present alternative modes of living beyond violent histories of extraction.

The techno-fix fantasies of Cixin Liu's 'Moonlight' explore the ways in which renewables cannot provide the sole answer to a just and ethical energy transition. The short story thus invites the question of how we can implement transition differently. Le Guin's *The Dispossessed* presents an ambiguous utopia in the world of Anarres, but it also demonstrates an ecological consciousness through shared suffering, one based on the unbuilding of walls, and on connecting with others. Le Guin's imaginary world of de-growth and anti-capitalist thought gestures towards different ways of engaging with our environments, offering alternative perspectives on energy sources outside capitalist regimes. Lem's *Solaris* begins with the colonial and capitalist venture, an Enlightenment project of science and discovery, to extract and master alien worlds. As the narrative develops, however, mastery faulters and non-masterful relations begin to emerge. Throughout my readings, an earthly vulnerable ethics emerges illuminated through the thought of Weil, Pick and Singh. Science fiction's exemplary contact with what is alien and Other offers a way of attunement to relations which may feel far and distant. The intricate patchwork I have forged of thinkers of vulnerability point us to a poetic and ethical environmental thought.

To return to the research collective After Oil, the science fiction narratives and the theoretical arc of this chapter offered ways to 'transform and transition our cultural and social values' (After Oil 2016, 10). The violent mechanisms and imperialist strategies that shape contemporary modes of energy extraction are critiqued by Cixin Liu's 'Moonlight', whereas *The Dispossessed* and *Solaris* provide a transformation that is non-anthropocentric, attentive to forms of oppression and suffering while presenting forms of solidarity across species lines to the alien and the unknown. The After Oil collective go on to note 'the energy question is, at its core, a human question, a social question that concerns accounting for the quality of human experience under the fossil economy, reckoning with the increasing precarity of life under fossil fuels' (After Oil 2016, 13).

The question I have intended to propose, however, is that the energy question is not simply 'at its core, a human question' (After Oil 2016, 13) but a question that engages all creatures, from plant and animal life. What if we change the centring of the human in energy discourse, and present instead the history and lives of creatures? How we engage and interact with energy must not be on the basis of human mastery but rather through the experience of vulnerable co-existence. As the After Oil Research collective have suggested, the energy impasse has no signposts—we have no map to find out where we go next. Perhaps then it is important to abandon what they call the humanist project of transition and instead invoke an earthly vulnerability. The science fiction narratives I have explored demonstrate the limits of human power and mastery, revealing what happens when mastery fails as a creaturely fellowship arises. A creaturely transition traces the logic of flesh, the vulnerability of all beings, discovering companionship with other creatures and the world around us. By advocating an energy transition not led by liberal humanism and its histories of violence but instead through dehumanist and creaturely solidarities, illuminating vulnerable earthly solidarities, new worlds become possible and a life beyond oil can be conceived.

References

After Oil. 2016. Edmonton, Alberta: Petrocultures Research Group.

After Oil. 2022. *Solarities: Seeking Energy Justice*. Forerunners: Ideas First Ser. Minneapolis: University of Minnesota Press.

Brown, Adrienne Maree. 2017. *Shaping Change, Changing Worlds*. Emergent Strategy Series, no. 0. Chico, CA; Edinburgh: AK Press.

Canavan, Gerry, and Kim Stanley Robinson, eds. 2014. *Green Planets: Ecology and Science Fiction*. Middletown: Wesleyan University Press.

Cha, Yoon Sook. 2017. *Decreation and the Ethical Bind: Simone Weil and the Claim of the Other*. New York: Fordham University Press.

Cock, Jacklyn. 2011. Green Capitalism or Environmental Justice. *Helen Suzman Foundation* 63 (Sustainability). https://admin.hsf.org.za/publications/focus/focus-63/Jacklyn%20Cock.pdf/view.

Jameson, Fredrid. 1996. *Seeds of Time*. Columbia University Press.

Jameson, Fredric. 2007. *Archaeologies of the Future: The Desire Called Utopia and Other Science Fictions*. Verso.

Jesson, Stuart. 2017. Compassion, Consolation, and the Sharing of Attention. In *Simone Weil and Continental Philosophy*. Rowan and Littlefield.

Jue, Melody. 2014. Churning Up the Depths: Nonhuman Ecologies of Metaphor in Solaris and 'Oceanic'. In *Green Planets: Ecology and Science Fiction*.

Le Guin, Ursula K. 2012. Forward. In *Roadside Picnic*. Gollancz.

Le Guin, Ursula K. 2014. *The Dispossessed*. First Perennial Classic Paperback edition, Reissued in Harper Perennial Modern Classics. Harper Perennial Modern Classics. New York, NY; London; Toronto; Sydney; New Delhi; Auckland: Harper Perennial.

Le Guin, Ursula K. 2018. *Dreams Must Explain Themselves and Other Essays, 1972–2004*. London: Gollancz.

Lem, Stanislaw. 2003. *Solaris*. Faber & Faber.

Liu, Cixin. 2019. Moonlight. In *Broken Stars*. Head of Zeus.

Macdonald, Graeme. 2016. Improbability Drives: The Energy of SF. *Strange Horizons*, no. February 15. http://strangehorizons.com/non-fiction/articles/improbability-drives-the-energy-of-sf/.

Malm, Andreas. 2020. *The Progress of This Storm: Nature and Society in a Warming World*. Paperback first published. London Brooklyn, NY: Verso.

Marder, Michael. 2017. *Energy Dreams: Of Actuality*. New York: Columbia University Press.

Shiva, Vandana et al. 2022. *Ecofeminism*. Critique Influence Change 05. London; New York; Dublin: Bloomsbury Academic.

Pick, Anat. 2011. *Creaturely Poetics: Animality and Vulnerability in Literature and Film*. New York: Columbia University Press.

Pick, Anat. 2017. Nothing Now but Kestrel': Simone Weil, Iris Murdoch and the Cinema of Letting Be. In *Iris Murdoch Review*. Kingston University Press.

Pick, Anat. 2018. Vulnerability. In *Critical Terms for Animal Studies*. Chicago University Press.

Prettyman, Gib. 2014. Daoism, Ecology and World Reduction in Le Guin's Utopian Fictions. In *Green Planets: Ecology and Science Fiction*. Wesleyan University Press.

Reid, Michelle. 2009. Postcolonialism. In *The Routledge Companion to Science Fiction*. Routledge.

Rieder, John. 2008. *Colonialism and the Emergence of Science Fiction*. The Wesleyan Early Classics of Science Fiction Series. Middletown, Conn: Wesleyan University Press.

Said, Edward. 1994. *Culture and Imperialism*. Vintage.

Singh, Julietta. 2018. *Unthinking Mastery: Dehumanism and Decolonial Entanglements*. Durham; London: Duke University Press.

Stillman, Peter, and Lawrence Davis. 2005. *The New Utopian Politics of Ursula K. Le Guin's The Dispossessed*. Lexington Books.

Szeman, Imre. 2014. Blind Faith. *Policy Options*. https://policyoptions.irpp.org/magazines/technology/blind-faith/.

Szeman, Imre. 2015. Entrepreneurship as the New Common Sense. *South Atlantic Quarterly* 114 (3): 471–490. https://doi.org/10.1215/00382876-3130701.

Vint, Sherryl. 2010. *Animal Alterity: Science Fiction and the Question of the Animal*. 1. publ. Liverpool Science Fiction Texts and Studies 39. Liverpool: Liverpool University Press.

Weheliye, Alexander G. 2014. *Habeas Viscus: Racializing Assemblages, Biopolitics, and Black Feminist Theories of the Human*. Durham: Duke University Press.

Weil, Simone. 2002. *Gravity and Grace*. Routledge.

Weil, Simone. 2009. *Waiting For God*. Harper Perennial Modern Classics.

Weil, Simone. 2021. *Simone Weil: An Anthology*. Penguin Classics.

Weinstone, Ann. 1994. Resisting Monsters: Notes on 'Solaris'. *Science Fiction Studies* 21 (2): 173–190.

Wilson, Sheena. 2018. Energy Imaginaries: Feminist and Decolonial Futures. In *Materialism and the Critique of Energy*. MCM Publishing.

Wynter, Sylvia. 2003. Unsettling the Coloniality of Being/Power/Truth/Freedom: Towards the Human, After Man, Its Overrepresentation--An Argument. *CR: The New Centennial Review* 3 (3): 257–337. https://doi.org/10.1353/ncr.2004.0015.

CHAPTER 6

Conclusion: Vulnerability, Hope and Energy Transition

In River Solomon's novel *The Deep*, the main character Yetu is a hybrid creature, part fish and part human. Yetu comes from a community of marine creatures raised by whales and whose ancestors were victims of the transatlantic slave trade, victims who were thrown overboard from a slave ship. In the passage that opens this conclusive chapter, Yetu reflects on a particularly human way of understanding the world, tied to violent classificatory and colonising systems which are about mastery and control rather than mutual co-existence:

> Two-legs had a specific way of classifying the world that Yetu didn't like. She remembered that, at least. They organised the world as two sides of a war, the two legs in conflict with everything else. The way Suka talked about farming was as if they ruled the land and what it produced as opposed to – they'd just said it themselves – existing alongside it. (Solomon 2020, 84)

Yetu's perspective aligns with my project in this book, of discovering common ties and bonds outside the exploitative relations of mastery and capital, allowing us to recognise vulnerability while existing mutually alongside other creatures and environments.

Towards the end of Solomon's narrative, we discover that the 'two legs', the 'surface dwellers', have ventured again into the depths of the ocean to search for and extract crude oil. As one of the characters describes,

J. Taylor, *The Nonhuman Narratives of Energy*, Palgrave Studies in Animals and Literature,
https://doi.org/10.1007/978-3-031-94332-4_6

'their purpose was to see what gifts of the deep they could steal from us. Below us, deep beneath the sand, there is a substance they crave. It is their life force. Their food. They feast on it like blood' (Solomon 2020, 135). Here, Solomon connects the trauma of the past, histories of colonisation and slavery, to the violence of the present, the regimes of extraction which destroy homes and dispossess communities. Solomon poignantly reflects on the power of multispecies communities in the face of violent regimes of extractive capital. In this extractive zone, deep within the ocean, these marine human hybrid creatures determine that 'we fight' (Solomon 2020, 136) against the oncoming destruction of their habitat.

By contemplating a world beyond the human, *The Deep* connects us to the vulnerability of different environments and bodies, of discovering multitudes, complexities and fellowship with other creatures. The ambitions of this book were to illustrate similar spaces of mutual and vulnerable co-existence which can emerge within extractive zones, illuminating the violence of extractive capital while also articulating voices of resistance. Exploring the coalescence of narratives energy and the nonhuman, I began first with an introductory example of the whale in *Moby-Dick*, the marine resource. Nineteenth-century whaling haunts present fossil practices and is a symbolic example of the ways the animal features in early energy production. The parallels I examined between meat and oil production showed the eery proximity of these two industries and the cognitive dissonance at the heart of their production. Moving to infrastructural violence of oil spills, roadkill and contaminated waters, there was a discovery of hope despite being mired by scenes of environmental ruin. Our final venture into the worlds of science fiction articulated limitations in technofixes to energy transition. *Solaris* and *The Dispossessed*, however, captured what is conceivable through an ethics of earthly vulnerability; they alert us to what might be possible and perhaps unblock the energy impasse and imagine a just multispecies future. Storying the nonhuman in extractive zones and histories of energy culture not only offers a magnification of our hidden present, the story of nonhuman in histories of energy, but may suggest ways beyond our present oily and extractive practices. The intricately woven voices of vulnerability from Butler, Weil, Pick and Singh demonstrate an environmental poetic in these works of art and culture, showing how companionship with who and what is far and seemingly distant can perhaps change our energy futures.

Gómez-Barris' exploration of extractive zones within regions of South America reveals what she coins 'submerged perspectives' and the 'unpredictability of the decolonial gesture' (2017, 13). Through entering into zones of production, land in which capital has rendered commodity, Gómez-Barris asks 'what are the more silent spaces that refuse these terms' (Gómez-Barris 2017, 14). The scenes of vulnerability in the chosen art and fictions offered these silent yet powerful modes of resistance. For Gómez-Barris, extractive zones contain seeds of resistance to the technologies which rendered them inert and passive. As Gómez-Barris suggests,

> Extractive zones contain within them submerged perspectives that challenge obliteration. I describe these transitional and intangible spaces as geographies that cannot be fully contained by ethnocentrism of speciesism, scientific objectification, or by extractive technologies that advanced oil fields, construct pipelines, divert and diminish rivers, or cave-in mountains through mining. Seeing and listening to these worlds present nonpath dependent alternative to capitalist and extractive valuation. (Gómez-Barris 2017, 12)

My readings of the cultural works explored paths of attunement to the vulnerability of different ecologies, acts of seeing and listening, turning to images and writing which illustrated capitalist violence yet generated voices and bodies of a hopeful solidarity across difference. An earthly vulnerability challenged the human mastery which is implicit in the ideology of extractive regimes. By informing energy ethics through the modality of vulnerability, the literary and cultural works offered different modalities of the human which has ramifications for energy relations.

As many petro-critics have noted, we cannot solely rely on green technologies and techno-fixes to solve the problems of the present; alongside a shift away from the fossil fuel economy, there must be a societal and cultural transition to challenge the current imperialist modes of energy production. My illustration of a planetary and earthly vulnerability, conveyed through cultural and literary works, serves to present a disruptive antithesis to the 'extractive view' and capitalist machinations which have shaped energy production. Energy is a problem for our current times, suggests Cara New Daggett,

> Humans need new energy systems – and likely new energy cultures—that leave fossil fuels in the ground and that instead rely on renewable fuels, coupled with more efficient technologies and, most likely, decreased energy consumption […] However, market based fixes are insufficiently appreciative of the limits of human mastery over the world. In order to live

appropriately on the Earth, humans need to re-evaluate our commitment to endless growth, productivity, and commodity accumulation. (Daggett 2019, 187)

Daggett highlights the limitations of green capitalism and dwells on the problems of relying on technological innovation to alleviate the issue of oil running dry while hoping to maintain endless growth and consumption that has defined the fossil fuel age. The philosophical strands developed in this book explore a different trajectory and marks a transition away from capitalist progress and human mastery. Through various conceptual tools of vulnerability, I present a non-extractive paradigm focused on vulnerability as a mode of exposure and attention. My work thus articulates the centrality of ethics and culture in order to shape our present relations to the land and other species, and to experiment with and explore the cultural imaginary which stages new ways of living with and attending to the Earth.

By outlining an alternative energy ethics on the basis of vulnerability, I present a politics of hope within the extractive zone and capitalist ruin. Combining the fields of petroculture and animal studies points towards a multispecies ethics which has implications for energy and energy regimes, demonstrating how it is possible to attune and to attend to vulnerable environments, and how to generate hope in extractive zones.

Oliva Laing's *Funny Weather: Art in an Emergency* reflects on the importance of hope in response to the emerging disasters of the twenty-first century. Laing quotes Eve Sedgewick's work as a way to forge different futures and contend with the present,

> Hope, often a fracturing, even traumatic thing to experience, is among the energies by which the reparatively positioned reader tries to organise the fragments and part-objects she encounters or creates. Because the reader has room to realise that the future may be different from the present, it is also possible for her to entertain such profoundly painful, profoundly relieving, ethically crucial possibilities as that the past, in turn, could have happened different from the way it actually did. (Sedgwick 2003, 146)

Laing focuses on what Sedgewick terms 'paranoid reading' and 'reparative reading', suggesting that this mode of reading is something that affects us all 'how we me make sense of the world, how we approach knowledge and uncertainty' (Laing 2021, 3). Most crucially, however,

Laing explores Sedgewick's impetus for hope. Writing at the times of AIDS crisis, Laing suggests Sedgewick's 'hope was hard-won, and in part derived from the powerful role art played during the plague years' (Laing 2021, 5). The energy transition is without significant clarity and direction, and we are met with the destabilising uncertainty Laing describes. Yet, as I have explored, an openness to vulnerable encounters offers room for hope, and where cultural and literary work can experiment with different futures while illuminating the problems of the present.

My chosen cultural and literary sources brought us to the interior of the extractive zone, magnifying the inequalities and violence which shape energy production. Literary and cultural expression offered alternatives to the present circumstances, generating hope and possibility of just and utopian energy futures. If, as petro-critics argue, the cultural imaginary is invested in fossil fuels, and we need to discover ways to divest, my work alternatively offered both an act of withdrawal and an intimacy with different creatures and environments. The exposure of vulnerability revealed not only how we are perishable and subject to harm but also how we form a powerful resistance to the present injustices which shape energy production. Interdependency, an essential component of an ethics of earthly vulnerability, generates bonds and connections which challenge the present hegemony of capitalism. It is through a shared vulnerability in which a multispecies solidarity emerges and alternative energy relations can be envisioned.

References

Daggett, Cara New. 2019. *The Birth of Energy: Fossil Fuels, Thermodynamics and the Politics of Work*. Durham, NC: Duke University Press.

Gómez-Barris, Macarena. 2017. *The Extractive Zone: Social Ecologies and Decolonial Perspectives*. Dissident Acts. Durham; London: Duke University Press.

Laing, Olivia. 2021. *Funny Weather: Art in an Emergency*. New York, NY: W.W. Norton & Company.

Sedgwick, Eve Kosofsky. 2003. *Touching Feeling: Affect, Pedagogy, Performativity*. Durham: Duke University.

Solomon, Rivers. 2020. *The Deep*. First Saga Press Paperback Edition. London; Sydney; New York; Toronto; New Delhi: Saga Press.

Index

J. Taylor, *The Nonhuman Narratives of Energy*, Palgrave Studies in Animals and Literature,
https://doi.org/10.1007/978-3-031-94332-4

The manufacturer's authorised representative in the EU is Springer Nature Customer Service Centre GmbH, Europaplatz 3, 69115 Heidelberg, Germany. If you have any concerns regarding our products, please contact ProductSafety@springernature.com

Printed and bound by CPI Group (UK) Ltd, Croydon, CR0 4YY

30/09/2026

02225514-0017